MYCOTOXIN DEACTIVATION

A SUCCESSFUL MYCOTOXIN TREATMENT AND REDUCTION CASE STUDY

DAVID MARK QUIGLEY

CONTENTS

FOREWORD

As a professional in the legal industry, I have had the distinct pleasure and opportunity to consult with some of the top names within the insurance industry on how to limit liability exposure for mold toxicity claims, reduce downtime illness for employees, and to create a safer atmosphere for anyone who might have had prolonged exposure to a building's interior air. Much focus was placed on the removal of the mold spores themselves in an effort to reduce costs to the carrier. However, the typical inspection, containment, filtration, removal system, despite its best intentions, was a flawed system. Typically, within weeks to months of what was believed to be the removal of the dangerous mold spores, rendering a building to be once again safe, reports of ongoing illness, allergies, and discomfort were routinely heard. This ongoing cycle caused significant impact to the carrier's bottom line due to additional remediation and inspection costs and, moreover, left many commercial landlords in the difficult position of facing considerable remedial costs to remove further contamination that was not necessarily located.

A second school of thought emerged that it wasn't the mold spore itself, per se, that was the culprit, but rather, the mycotoxins naturally produced by mold that is causing illness. Unlike the mold spore, the mycotoxin is not alive. Therefore, it needs to be broken down, not killed like a mold spore. The difficulty with current clean-up methods is that while they may serve to remove the mold spore, they routinely further disperse the mycotoxin throughout an enclosed building. Until recently, notwithstanding complete demolition and removal, the two manners most effective for complete removal of these dangerous mycotoxins were either superheating a room to over 500 degrees Fahrenheit, which is simply impractical or, to deactivate the mycotoxin with enough ozone that would prove fatal to humans.

Part of risk management is to stay current on industry technology that has been proven beneficial by independent lab analysis. As the mycotoxin deactivation sciences are still in their relative infancy, a RealTime Labs white paper piqued my interest that complemented a case study claiming quantifiable mycotoxin deactivation using manners that were not lethal to humans or that required substantial structure downtime. Being a skeptic, I reviewed both the white paper and case study which seemed to present well enough. But, too often, white papers as well as case studies make bold claims, only to grossly disappoint in real-world application.

I had the opportunity to follow a live field application of the BioRisk mycotoxin decontamination protocol to see if it did, in fact, hold up to the claims. Surface sampling was conducted in accordance with ELISA technology; accepted and approved by the USDA for mycotoxin evaluation standards, and pre-testing sample collection went beyond normal testing areas to include sample collection from multiple surface types, exposures, and airflow areas. All sites were made available to those in attendance to ensure testing was legitimate. The application of the BioRisk Decontamination process was surprisingly quick by the hygienist, with minimal downtime associated with the application process, and the building was immediately occupiable upon conclusion of the post-testing sample validation. Surprisingly, the post-collection samples showed substantial mycotoxin deactivation, and the decrease in mycotoxins was considerable – showing total reduction – when compared to other removal methods. Further, the downtime of the building was vastly reduced when compared to traditional mycotoxin deactivation methods.

Most importantly, deactivation occurred in hidden areas where it would not be expected, contrary to the results of standard mycotoxin removal, which left active particles behind.

We have seen these same results repeated time and again by BioRisk. Their technique is clearly ahead of its time and should be considered industry standard. The application of the BioRisk protocol limits downtime, is cost effective, and deactivates substantially far greater mycotoxin numbers, while being safe to humans. It is inevitable that the use of this technique over other mycotoxin clean-up protocols, in conjunction with mold removal, will significantly impact the bottom line for any risk manager. The BioRisk process is a game-changer for the building industry without question. I could not recommend any mold removal/remediation method without demanding the inclusion of a BioRisk application for mycotoxin deactivation.

James Moon - Attorney

(James Moon is a legal consultant to the insurance industry and is recognized as an expert witness on loss-related claims in federal court. He is licensed and actively practices in New York, Michigan, Florida, and Tennessee. Additionally, he is certified as a Risk Analyst through the Global Academy of Finance and Management. He has participated as lead counsel in over 1,000 commercial and multi-family residential loss claims involving mold loss and mycotoxin removal.)

EXECUTIVE SUMMARY

This project specifically focuses on the successful deactivation of detected mycotoxins.[1]

Risk Assessment
- Undertake a site-specific bio recovery risk assessment to establish potential and real pathogenic, biosafety, and life safety risks associated with this project.

Establish Extent of Contamination
- Identify if mycotoxins are present within the client's office and/or HVAC System via surface sample analysis by an accredited third-party laboratory.
- If it is established that mycotoxins are present, deactivation of the said mycotoxins will be undertaken.

Sampling and Deactivation Methodology
- The laboratory will undertake an immunological test to detect if mycotoxins are present. If the antibodies used in the test match and bind to the mycotoxins in question, this reaction shows that they are present.
- Deactivate detected mycotoxins via an ionized H_2O_2 (hydrogen peroxide) deactivation treatment.

---

1   This Case Study in no way negates the necessity for appropriate and timely mold remediation when the presence of mold has been identified. The study's objective, using the author's QMD Protocol[SM], is to focus specifically on the assessment, identification, and validated deactivation of mycotoxins.

- After undertaking the ionized H_2O_2 deactivation treatment, establish if mycotoxins have been deactivated, with any corresponding reductions documented via validation surface sample analysis by an accredited third-party laboratory.

Anticipated Results

- The ionized H_2O_2 treatment, outlined in this report, is intended to deactivate the mycotoxins and render them inert. The radical oxygen species, the hydroxyl radical, produced by the process splits or breaks the chemical molecular structure of the mycotoxins.
- Using the same immunological test as used for identification purposes, show that the ionized H_2O_2 deactivation treatment successfully splits or breaks the intramolecular structure of the previously detected mycotoxins, and the antibodies used in the test are no longer able to bind to a matching mycotoxin. Doing so validates the molecular structure of the mycotoxins previously detected no longer exist, having been deactivated and rendered inert.

Deactivation

- As the presence of mycotoxins was established during the identification phase, the client's office – specifically the HVAC System – was biohazard treated using an ionized H_2O_2 treatment as outlined in this report.

Results

- Via the use of validation sample analysis by an accredited third-party laboratory, the efficacy and effectiveness of the ionized H_2O_2 deactivation treatment has been established and produced a total reduction in mycotoxin contamination.

  - ❖ Using the ionized H_2O_2 treatment, as outlined in this report, to deactivate mycotoxins, it has been established by validation sampling that the radical oxygen species produced by the process split or broke the molecular structure of the mycotoxins. Doing so shows that the previously detected molecular structure of the mycotoxins no longer exists, thus successfully deactivating them and rendering them inert.

- The QMD Protocol^{SM}, covered in-depth later, was highly successful and effective. The Assessment, Identification, and Deactivation Stages achieved their objectives, and Validation showed that all detected mycotoxins were deactivated and rendered inert.

Want more information about this process?
To contact the author for biohazard sampling, consulting, and deactivation services, or book him as a keynote speaker,
email info@biorisk.us,
call (877) BioRisk,
or visit www.biorisk.us/mycotoxin_deactivation

INDEPENDENT CASE STUDY REVIEW

"The White Paper on Mycotoxins is a qualitative study reporting the documented results of Pre-Validation Sampling regarding the levels of detectable mycotoxins and mold in an HVAC system.

"Following is the Decontamination Protocol and the Biohazard Disinfection process. Next follows the Post-Validation Sampling which provides quantifiable results regarding the reduction of mycotoxin and mold with the HVAC system.

"The purpose of this study is laid out cohesively and the table of contents is listed in a structured and searchable manner. The data found within this study is explained within the correct subheadings but also backed up with photographic evidence as the White Paper explains its findings.

"This study is conclusive with its findings and explains the importance of said findings."

ML, President & CEO – Leading and Respected International Infection Control Risk Assessment (ICRA) and Healthcare Consultant[2]

---

2   Reviewer requested anonymity due to governance requirements of a Board of Trustee on which they sit.

AUTHOR PROFILE
&
MYCOTOXIN EXPERIENCE

Although this case study focuses on my experience deactivating mycotoxins in the United States, my journey began in New Zealand nearly a decade ago.

As a licensed Asbestos Assessor and Removalist, previously holding a Certificate of Competency to undertake restricted work involving asbestos, my consulting company had taken over two-hundred-thousand samples and many thousands of surveys and reports.

I first encountered mycotoxins on a routine asbestos sampling job in 2013. The government agency that commissioned the job also requested mold sampling. Because I was coordinating and overseeing the work, I was only on-site for a few minutes, but I remember noticing the distinct musty smell of the building as soon as I entered. I was the only person present not wearing respiratory equipment.

That evening I began to feel unwell, and the next day I was bedridden. The day after, feeling extremely ill, I had to fly home to Florida. I was mostly incapacitated for the next six weeks and under the care of a leading doctor specializing in Microbiology, Allergy and Immunology, and Immunoparasitology.

After extensive testing, they discovered that I was severely allergic to mold, and my brief time in the building had caused widespread internal contamination.

My doctor prescribed a personalized and proprietary treatment plan (similar to the Shoemaker Protocol), and I recovered slowly over the next several months. My illness gave me a healthy respect for the serious effects mold contamination can have on the human body.

Due to my extreme allergy, I have chosen not to pursue a license in the subject. However, my experience led me to conduct extensive investigations of mold, and more particularly, its by-products, mycotoxins, toxicity, and potential ways to counteract and deactivate them, leading me, instead, towards biotoxins and the biohazard field.

It is my firm belief that being severely dyslexic aided my pioneering discovery, Mycotoxin Biodeactivation[SM], and establishment of the QMD Protocol[SM] and its **Q**uantitative **M**ycotoxin **D**eactivation[SM]. It wasn't until my late twenties that I was finally diagnosed with dyslexia, and with this came the realization that I think differently than those without the condition. This blessing, and my previous contamination, have allowed me to come at the formerly vexing problem of mycotoxin deactivation from an entirely different direction and thus gave me the ability to solve it.

Want to contact the author about biohazard sampling, consulting, and deactivation services, or book him as a keynote speaker?
email info@biorisk.us, call (877) BioRisk,
or visit www.biorisk.us/mycotoxin_deactivation

FROM CONTAMINATION TO DEACTIVATION

One of the difficulties the author discovered with mycotoxin contamination is their effective deactivation and challenging long-held beliefs that it cannot be done in situ. The old solution is to completely remove and replace all building materials, fabrics, and mechanical equipment that have been contaminated.

In mid-2020, a group of Florida's environmental experts invited the author to be queried about the extent of his asbestos experience. During this meeting, he was asked, due to his biohazard experience, if he believed he could deactivate mycotoxins without the costly exercise of removal.

Having always been a creative problem solver, and now armed with extensive research and first-hand knowledge of the effects of mycotoxin exposure, this meeting and its fateful question set the wheels in motion to effectively think-outside-the-box, implement a documented process, and successfully deactivate mycotoxins while leaving the environment in place.

What he achieved is a game-changer for the infectious control and biohazard remediation industries.

The following outlines the author's QMD Protocol[SM].

The QMD Protocol^{SM}

Though **QMD** are the author's initials backward (**D**avid **M**ark **Q**uigley), and he developed and reverse-engineered the QMD Protocol^{SM}, the initialism refers to **Q**uantitative **M**ycotoxin **D**eactivation^{SM}. The QMD Protocol^{SM} and its **Q**uantitative **M**ycotoxin **D**eactivation^{SM} focus on Mycotoxin Biodeactivation^{SM} and address some of the most toxic compounds known to humans. The Assessment, Identification, Treatment, and Validation stages make up the complete process of ridding interior environments of the deadly toxigenic biological hazard of mycotoxins.

The QMD Protocol^{SM} focuses on the low molecular mass-secondary metabolites that mold produces. In terms of human health, it's not the mold we should be most worried about; it's the toxins it produces, namely mycotoxins.

The QMD Protocol^{SM} has four critical components or phases, each equally important as the others.

1. **Assessment**

   A risk assessment of the environment suspected to be contaminated needs to be undertaken first. This is done from a biohazard standpoint, adapted for but not limited to mycotoxin contamination, using bio recovery site risk assessment, and the author's knowledge of CFHC ICRA (Critical Facilities Healthcare Infection Control Risk Assessment) guidelines[3].

---

3   An example of the bio recovery risk assessment and its guidelines can be found here: Here

The following factors are used in the assessment:

- Determine the type of risks present – what is the Bio Safety Risk Group they fall into as recommended by ABSA Risk Group Data Base.

- Identify the Area Risk Group Classification – either offering Low, Medium, or High impact risk levels of the project's site/area (including surrounding areas).

- Determine Class (I – IV) of Risk Mitigation Measures Required (calculated by measuring the Risk Group against the Area Risk Group impact level).

- Risk Mitigation Guidelines (established by Class I, II, III, or IV risk mitigation measures). This will determine workflow and required PPE used to mitigate risk during the project.

- Life Safety Assessment (a nine-step assessment as outlined by Life Safety Code and building code requirements).

- Risk Assessment Sign-Off (sign-off by the client, project manager, and technicians involved in the project).

Each potentially contaminated property is made up of micro and macro environments; essentially, it is a living environment unto itself and needs to be assessed as such.

Ideally, a visual inspection would be undertaken during this phase. However, it could also be undertaken during the identification phase.

2. Identification

It is not only important to identify the presence (or absence) of mycotoxins, but also to establish in-depth quantitative reporting of their type and quantity.

Methodical sampling procedures need to be followed to ensure accurate surface samples are taken, with their analysis being undertaken and reported by a reputable independent third-party laboratory.

3. Treatment

The logic of an effective deactivation treatment may, on the surface, appear to speak for itself. However, mycotoxins have shown themselves to be very difficult to efficaciously deactivate in situ, with even successful mold remediations still leaving them in place.

The current treatment paradigm was the costly exercise of removing all potentially contaminated building materials, surfaces, fabrics, and associated mechanical equipment. Instead, the solution arises upon consideration of the contamination as a biological, chemical, and mathematical problem. Then, using all three of these scientific disciplines to solve it synergically.

Although mycotoxins are chemical compounds, they are considered biotoxins, because they are produced from mold, a biological substance. Considering this, the treatment component of the QMD Protocol[SM] looks at breaking the molecular structure of the identified mycotoxins. All that remains for effective treatment is to then calculate the affected area and apply the desired volume of treatment to it. Once this is achieved, their chemical makeup is broken, rendering them inert, efficaciously treating the problem.

If any gross bio-contaminations are identified, such as mold, bacteria, or other pathogenic and indoor pollutants, during the assessment or identification phases, alternative remediation protocols may be required before the QMD Protocol[SM] treatment is undertaken.

Additionally, if it is determined that an extensive mycotoxin contamination is present during the identification phase, load reduction may be used to balance and reduce the load before the QMD Protocol[SM] treatment is undertaken.

4. **Validation**

It is all very well assessing the risks of a situation, identifying that a mycotoxin contamination exists, and undertaking an effective deactivation treatment program; however, validation – determining that mycotoxins have been rendered inert – is vital.

As with the initial identification, surface sampling of the affected areas needs to be accurately gathered, and their analysis again needs to be undertaken by an independent reputable third party to show the desired results have been achieved, thus validating them.

The QMD Protocol[SM] Summary
The above describes the complete QMD Protocol[SM]. When all of these components are undertaken chronologically and methodically, and exactingly executed, the desired outcome is assured.

The QMD Protocol[SM], and its **Q**uantitative **M**ycotoxin **D**eactivation[SM] is a sophisticated development, complementary to Mycotoxin Biodeactivation[SM]. This innovative approach is a revolutionary advancement and has been considered from biohazard and bio-recovery standpoints.

The QMD Protocol[SM] provides a serious deactivation solution for multiple environments and industries. When adopted, fields ranging from food and coffee processing, animal feed production, and deactivation of weaponized aflatoxins, to name a few, will benefit enormously.

How the site-specific QMD Protocol[SM] was implemented in this case study is described in further detail in the following pages.

Consult Team & Parties Involved

The author's biohazard consulting company, BioRisk Decontamination & Restoration, LLC (BioRisk), has been commissioned by International Construction Partners, LLC (ICP and the Client), using the QMD Protocol[SM], to assess, identify, and show validated deactivation of any mycotoxin within their office and its HVAC system.

Consulting Personnel Required

Although mycotoxins are chemical compounds, they are considered biological hazards instead of chemical hazards because their presence is associated with fungal contamination occurring at some point in the associated environment.

Therefore, biohazard expertise and a proven deactivation decorum, the QMD Protocol[SM], are required to assess, identify, treat and deactivate mycotoxin contamination effectively.

Biohazard Consultant

- BioRisk is permitted and registered with the Florida Department of Health for Biohazard Cleanup and Transportation, DOH 11-64-2049521, DOH 11-64-2049525.
- Although mycotoxins are recognized as a biohazard, Mr. Quigley of BioRisk carries OSHA 40 Hour HAZWOPER 1910.120 (e) pursuant to 29 CFR 1910.120 OSHA HAZWOPER Regulations.
- BioRisk has been engaged in a consultancy capacity only for the above, in relationship to the Scope of Work beneath, and within the limitations outlined in this report's Limitations and Disclaimers.
- BioRisk is also certified for the use and environmental application of ionized H_2O_2.

SCOPE OF WORK

This project specifically focuses on the risk assessment, identification, and validated deactivation of mycotoxins.

The scope of work using the QMD Protocol^{SM} is as follows:
- Undertake a site-specific bio recovery risk assessment to establish any potential and real risks associated with client's office and/or HVAC System.
- Identify if mycotoxins are present within the client's office and HVAC System via surface sample analysis by an accredited third-party laboratory.
- As active mold was evident at the HVAC Returns, and specifically within the HVAC Air Handler (see Site Images), during the initial inspection, the office environment and, in particular, the HVAC System are to be decontaminated, and biohazard treatment applied using an ionized H_2O_2 treatment protocol.
- Establish if any identified mycotoxins have been deactivated, with any corresponding reductions determined via validation surface sample analysis by an accredited third-party laboratory.

Scope of Work Specifics

Site Address:	2710 Brantley Blvd, Naples, Florida 34117
Report Commissioned by (Client):	International Construction Partners, LLC
Type of Report:	Mycotoxin Assessment, Identification & Deactivation
QMD Protocol[SM]:	Quantitative Mycotoxin Deactivation[SM]
Assessment:	Bio Recovery Site Risk Assessment Guidelines
Identification:	Lift, Swab & Gauze Surface Sampling - EMMA (Environmental Mold & Mycotoxin Assessment)
Treatment:	ionized H_2O_2
Validation:	Lift, Swab & Gauze Surface Sampling - EMMA (Environmental Mold & Mycotoxin Assessment)
Validating Laboratory	RealTime Laboratories, INC, 4100 Fairway Dr Ste 600, Carrollton, TX 75010
Assessment Date:	September 12th 2020
Identification Date:	September 12th 2020

Treatment Date:	September 16th 2020
Validation Sampling Date:	September 16th 2020
QMD ProtocolSM Assessment, Identification, and Treatment, Reporting, and Biohazard Consultation:	David Mark Quigley, BioRisk Decontamination & Restoration, LLC©, FDOH 11-64-2049521, & 11-64-2049525
Date Report Issued:	October 15th 2020
Author:	David Mark Quigley©
Author's Signature:	

Site Location

This report presents the documented results of the following:
- Assessment: a site-specific bio recovery risk assessment of potential and real risks.
- Identification Surface Sampling: the levels of detectible mold and mycotoxins within the HVAC system.
- The Managed Decontamination and Biohazard Treatment used to deactivate the mycotoxins present within the HVAC system.
- Validation Surface Sampling: the quantifiable reduction of mold and mycotoxin from within the HVAC system.

Reduction Summary
Please see report sections titled Mold & Mycotoxin Reduction and Conclusions, respectively, which quantify the successful decontamination and subsequent reduction of all detected mycotoxins using the above outlined strategy and the prescribed Scope of Work.

Declaration

The author declares that the material of this report is his own work and relating to the project herein being documented.

At the time of conducting this case study, the author is a licensed and certified Asbestos Assessor in New Zealand and Australia, and certified in the USA for the following: as a Hazardous Waste Generator pursuant to RCRA 40 CFR 262.34, 264.16, 265.16; in Bloodborne Pathogens in commercial and industrial facilities pursuant to OSHA 29 CFR 1910.11030; in Personal Protective Equipment (PPE) pursuant to OSHA 29 CFR 1910.132; Biosecurity trained pursuant to DOT HM-232, and DOT 49 CFR 174.704(a)(4), (a)(5), and DOT 49 CFR 172.802, via Security Plan Requirements DOT 49 CFR 172.800; 802, and OSHA 40 Hour HAZWOPER 1910.120 (e) pursuant to 29 CFR 1910.120 OSHA HAZWOPER Regulations. He is also a GBAC STAR™ Certified Technician with emphasis on SARS-CoV-2.

Additionally, BioRisk is permitted and registered with the Florida Department of Health for Biohazard Cleanup and Transportation, DOH 11-64-2049521, DOH 11-64-2049525, is certified for the use and application of ionized H_2O_2 and is a related entity to ICP.

Further, at the time of publication, the author carries accreditation for CFHC ICRA (Critical Facilities Healthcare Infection Control Risk Assessment) meeting requirements for the Environment of Care and the built environment - "Fundamentals of Infection Control for the Healthcare Contractor".

Want more information?
Email info@biorisk.us, call (877) BioRisk,
or visit www.biorisk.us/mycotoxin_deactivation

BACKGROUND

Why Target Mycotoxins

Many scientists, clinicians, and medical professionals state: "It's not the mold you should be most worried about…it's the toxins they produce."[4]

This report and the QMD Protocol[SM] covered herein focuses on the toxins – low molecular mass secondary metabolites – that mold produces, namely mycotoxins.

Mycotoxins are associated with mycotoxicosis, which can cause diseases and death in humans and animals and are well documented for their toxic effects on human cells. They are associated with numerous problems in normal cellular function and with a wide variety of clinical symptoms and diseases, such as:

- Kidney Toxicity
- Immune Suppression
- Autism
- Neurotoxicity
- Depression
- Aplastic Anemia
- Congenital Disorders
- Chronic Fatigue Syndrome
- Cancer
- Acute Pulmonary Hemorrhage

---

4   *Emma Environmental Mold and Mycotoxin Assessment & Testing.* RealTime Labs. (2021, October 3). Retrieved February 8, 2022, from https://realtimelab.com/emma/

EMMA Test Analysis

The EMMA test uses sensitive molecular detection technology to establish the presence and relative abundance of 10 of the most toxigenic molds. It also tests for 16 of the most poisonous mycotoxins.

Please see report sections titled EMMA Mold Test Panel and EMMA Mycotoxin Test Panel for specific information on what EMMA tests for.

Redundancy Sampling

Due to the critical nature of this report, the restricted quantities of sample material, the time-dependent window of available identification surface sampling material, the client requiring immediate balancing of the office environment and treatment of the HVAC System after identification surface samples are taken, a redundancy sampling protocol will be used. A batch of secondary samples of all sample locations will be taken and held in reserve in case alternative sample analysis is required.

Office Layout & HVAC System

The client's office and warehousing complex consists of two offices, a half bathroom, and two open warehousing and vehicle bays accessed by metal roller doors. The two offices and half bathroom are serviced by a 1 ½ ton HVAC system. The air return and supply consist of the following:

- Two eight-inch ceiling and two ten-inch floor returns
- Three eight-inch air supplies per office
- One six-inch ceiling supply in the half bathroom

One of the two offices was originally designed and built as a "wet room," complete with an open floor drain, but was never used as such.

Category 3 Water Contamination

The client's office is not connected to the city sewer. It has a standalone septic system, which is shared with an adjacent building. In July 2020, the office complex and surrounding area experienced severe thunderstorms and lightning activity. On a number of occasions, lightning strikes disabled the septic system's electrical circuitry, preventing the system's lift station from pumping effluent from its holding tanks to its drain field.

In July 2020, during one of the periods when the office's septic system's electrical circuitry was disabled by a lightning strike, Category 3 water backed up through the building's grey and black water lines then passed through the P-trap of the office's open floor drain, pooling for several hours on the office floor until its discovery. Once the septic system's electrical circuitry was reactivated, the system's lift station pumped the pooling water away. The area was subsequently cleaned and disinfected.

Mold & Mycotoxin Contamination

During August 2020, staining was noticed in the office ceiling around the HVAC Air Supplies, and mold growth was noted on the HVAC Air Returns. Due to this, the HVAC Air Handler was inspected and opened under supervision, and active mold growth was noted within the mechanism and on the Air Handler's Core. (See Site Images.)

After the initial inspection, and when the client's staff members using the office began experiencing respiratory and health issues, the office was vacated and BioRisk was engaged to:

1. Undertake a site-specific bio recovery risk assessment.
2. Undertake identification surface sampling to determine the extent of any toxic mold and mycotoxin contamination.
3. Immediately deactivate any mycotoxin within the HVAC system.
4. Undertake validation sampling to determine if the reduction of mycotoxins had been achieved.

At the time of writing, due to the extent of mycotoxin contamination and severity of the health issues experienced by the client's staff members, under medical consultation, they have chosen to undertake the Shoemaker Mold Detox Protocol to detox their bodies from the effects of the mycotoxins.

Other Contributing Factors

Mold and mycotoxin require a number of conditions to grow and spread. All of the following conditions were present within the client's office:

1. Organic food sources, especially cellulose (e.g., paper, cardboard), present via floor return filters, adjacent to open floor drain. (See Site Images.)
2. An open water source. (See Site Images.)
3. Humidity ($\geq$67% RH). (See Site Images.)
4. Moderate temperature ($\geq$68-86°F). (See Site Images.)
5. Time of exposed Category 3 water contamination – several hours.

Site Images

Open Water Source

Proximity of Floor Return
Filter to Open Drain

Relative Humidity @ 77.3

Temperature @ 78.2 °F

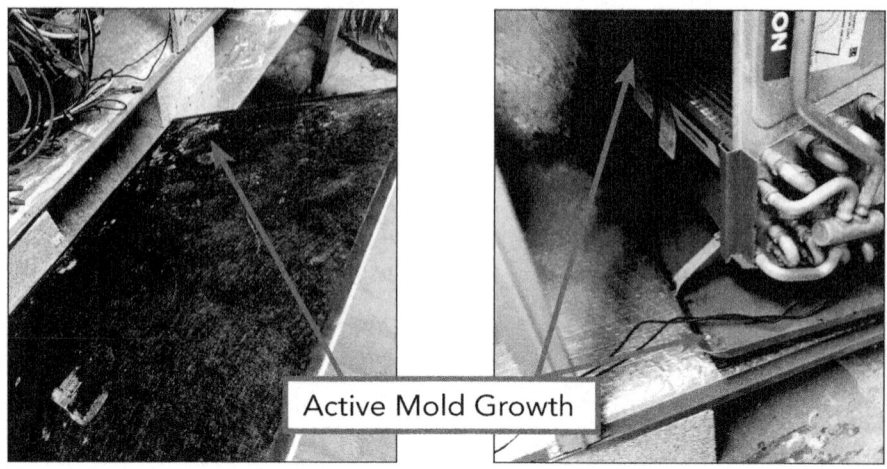

Active Mold Growth – within
the Air Handler

Active Mold Growth –
Core Tray and Core

Active Mold Growth – within
the Air Handler

Active Mold Growth –
Floor Return

Interested in learning more?
Email info@biorisk.us, call (877) BioRisk,
or visit www.biorisk.us/mycotoxin_deactivation

QMD PROTOCOL[SM] ASSESSMENT[5]

During the initial information intake and visual inspection of the site, the following summary of factors for the bio recovery site risk assessment were determined for this mycotoxin deactivation and the QMD Protocol[SM]:

- Risk Group – 2
- Area Risk Group – Medium
- Risk Mitigation Measures – Risk Group 2 / Area Risk Group Medium
- Risk Mitigation Guideline – Class II
- Life Safety Assessment – Passed
- Complete Risk Assessment – Signed off by all parties

---

5   Due to confidentiality and HIPAA Compliance the full risk assessment conducted is not shared in this case study.

IDENTIFICATION & VALIDATION SAMPLING

- This report and the QMD Protocol[SM] consist of a visual inspection, including surface sampling and third-party laboratory analysis of the material.
- All materials are analyzed by a CAP and CLIA accredited laboratory, regulated to perform Clinical Mycotoxin Testing.
- Samples that may contain mold and mycotoxins are analyzed by the ELISA technology; accepted and approved by the USDA for mycotoxin evaluation, using monoclonal or polyclonal antibodies for detection.
- The laboratory's test panel tests for the presence of ten toxigenic molds and sixteen of the most common mycotoxins, including ten macrocyclic Trichothecenes and the Class 1 carcinogen mycotoxin, Aflatoxin B1.
- Please see the next page for the interpretation guide of the Environmental Mold and Mycotoxin Assessment (EMMA) testing panel that has been used for the identification and validation surface sampling in this case study[6].

---

6   Please see Appendices for brief explanation guides on, environmental mold panel testing, mycotoxin panel, EMMA environmental test specifications, and fungal load explanation.

EMMA Interpretation Guide

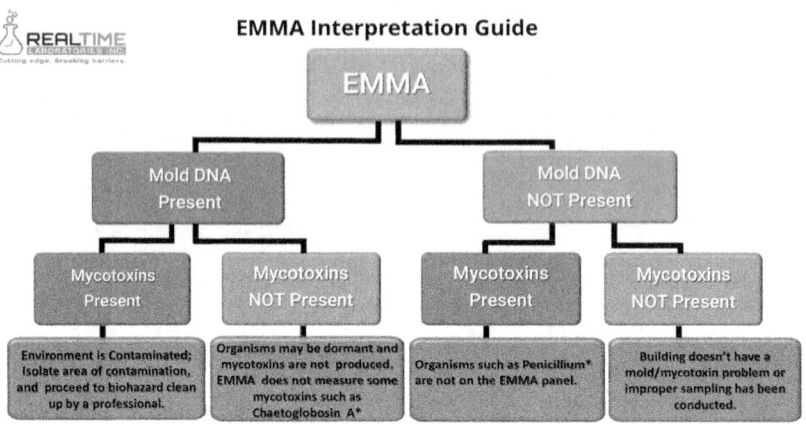

*EMMA testing panel does include organisms that are proven to cause serious illness and disease

QMD PROTOCOL[SM] IDENTIFICATION – LOCATION AND IDENTIFICATION OF MOLD & MYCOTOXINS

Location Description	ICP Office	ICP Office	ICP Office
Location of Sample	HVAC Air Return	HVAC Air Handler	HVAC Air Supply
Sample Description	Composite: Floor Return Filter, Floor Return Ducting, Ceiling Return Filter, Ceiling Return Ducting	Composite: Cover Insulation, Core Drain Tray, Core Drain Tray Insulation	Composite: Supply Plenum, Supply Vent 1, Supply Vent 2
Sample Number	PREV 1	PREV 2	PREV 3
Laboratory ID	PREV 1	PREV 2	PREV 3
Type of Mold	Aspergillus fumigatus (0.4927ng of DNA/mL) Aspergillus niger (13.3148ng of DNA/mL) Chaetomium globosum (0.03ng of DNA/mL)	Aspergillus niger (3.7282ng of DNA/mL)	No EMMA Mold Detected
Type of Mycotoxin	Ochratoxin A (12.62ppb) Aflatoxin Group (1.727ppb) Trichothecene Group (Macrocyclic) (0.042ppb) Gliotoxin Derivative (0.868ppb)	Aflatoxin Group (1.283ppb) Gliotoxin Derivative (9.884ppb)	No EMMA Mycotoxins Present
Pass/Fail	Fail	Fail	Pass

Site Comments
- Unless otherwise stated, all areas within the Scope of Work were accessible.
- Mycotoxin contamination was detected in substantial concentrations within the HVAC Air Return system and Air Handler mechanism. This leads to the conclusion that the environment that it draws from is additionally contaminated. Decontamination and a biohazard treatment were required for it and the entire HVAC System.

Important Notes
- This document sets out to identify the type, location, and description of any mold and mycotoxins identified pursuant to this report.
- This report should be read in conjunction with all sections, in particular the Detailed Identification Sample Results and the Detailed Validation Sample Results (which may contain detailed information and additional notes specific for that sample location), and specifically with reference to the associated laboratory reports accompanying this report.

DETAILED IDENTIFICATION
SAMPLE RESULTS

Sample Number	PREV 1
Location Description	ICP Office
Location of Sample	HVAC Air Return
Sample Description	Floor Return Filter
Laboratory ID	PREV 1
Mold Type	Aspergillus fumigatus Aspergillus niger Chaetomium globosum
Mycotoxin Type	Ochratoxin A Aflatoxin Group: B1, B2, G1, G3 Trichothecene Group (Macrocyclic): Roridin A, Roridin E, Roridin H, Roridin I-2, Verrucarin A, Verrucarin J, Satratoxin G, Satratoxin H, Isoatratoxin F Gliotoxin Derivative
Notes / Comments:	Composite Sample. Filter heavily impacted with dust – live mold evident. As mold and mycotoxins detected, Validation Sampling will be required.

Sample Number	PREV 1
Location Description	ICP Office
Location of Sample	HVAC Air Return
Sample Description	Floor Return Ducting
Laboratory ID	PREV 1
Mold Type	Aspergillus fumigatus Aspergillus niger Chaetomium globosum
Mycotoxin Type	Ochratoxin A Aflatoxin Group: B1, B2, G1, G3 Trichothecene Group (Macrocyclic): Roridin A, Roridin E, Roridin H, Roridin I-2, Verrucarin A, Verrucarin J, Satratoxin G, Satratoxin H, Isoatratoxin F Gliotoxin Derivative
Notes / Comments:	Composite Sample. As mold and mycotoxins detected, Validation Sampling will be required.

Sample Number	PREV 1
Location Description	ICP Office
Location of Sample	HVAC Air Return
Sample Description	Ceiling Return Filter
Laboratory ID	PREV 1
Mold Type	Aspergillus fumigatus Aspergillus niger Chaetomium globosum
Mycotoxin Type	Ochratoxin A Aflatoxin Group: B1, B2, G1, G3 Trichothecene Group (Macrocyclic): Roridin A, Roridin E, Roridin H, Roridin I-2, Verrucarin A, Verrucarin J, Satratoxin G, Satratoxin H, Isoatratoxin F Gliotoxin Derivative
Notes / Comments:	Composite Sample. Filter heavily impacted with dust – live mold evident. As mold and mycotoxins detected, Validation Sampling will be required.

Sample Number	PREV 1
Location Description	ICP Office
Location of Sample	HVAC Air Return
Sample Description	Ceiling Return Ducting
Laboratory ID	PREV 1
Mold Type	Aspergillus fumigatus Aspergillus niger Chaetomium globosum
Mycotoxin Type	Ochratoxin A Aflatoxin Group: B1, B2, G1, G3 Trichothecene Group (Macrocyclic): Roridin A, Roridin E, Roridin H, Roridin I-2, Verrucarin A, Verrucarin J, Satratoxin G, Satratoxin H, Isoatratoxin F Gliotoxin Derivative
Notes / Comments:	Composite Sample. As mold and mycotoxins detected, Validation Sampling will be required.

Sample Number	PREV 1
Location Description	ICP Office
Location of Sample	HVAC Air Return
Sample Description	Return Plenum
Laboratory ID	PREV 1
Mold Type	Aspergillus fumigatus Aspergillus niger Chaetomium globosum
Mycotoxin Type	Ochratoxin A Aflatoxin Group: B1, B2, G1, G3 Trichothecene Group (Macrocyclic): Roridin A, Roridin E, Roridin H, Roridin I-2, Verrucarin A, Verrucarin J, Satratoxin G, Satratoxin H, Isoatratoxin F Gliotoxin Derivative
Notes / Comments:	Composite Sample. As mold and mycotoxins detected, Validation Sampling will be required.

Sample Number	PREV 2
Location Description	ICP Office
Location of Sample	HVAC Air Handler
Sample Description	Cover Insulation
Laboratory ID	PREV 2
Mold Type	Aspergillus niger
Mycotoxin Type	Aflatoxin Group: B1, B2, G1, G3 Gliotoxin Derivative
Notes / Comments:	Composite Sample. Live mold growth evident. Results reported from redundancy samples[7]. As mold and mycotoxins detected, Validation Sampling will be required.

---

7   Sample analysis reported from redundancy sample material. Original results inconclusive from active mold growth sampled, as well as reported results inadvertently mixed with results from a non-client location.

Sample Number	PREV 2
Location Description	ICP Office
Location of Sample	HVAC Air Handler
Sample Description	Core Drain Tray
Laboratory ID	PREV 2
Mold Type	Aspergillus niger
Mycotoxin Type	Aflatoxin Group: B1, B2, G1, G3 Gliotoxin Derivative
Notes / Comments:	Composite Sample. Live mold growth evident. Results reported from redundancy samples[8]. As mold and mycotoxins detected. Validation Sampling will be required.

8   Sample analysis reported from redundancy sample material. Original results inconclusive from active mold growth sampled, as well as reported results inadvertently mixed with results from a non-client location.

Sample Number	PREV 2
Location Description	ICP Office
Location of Sample	HVAC Air Handler
Sample Description	Core Drain Tray Insulation
Laboratory ID	PREV 2
Mold Type	Aspergillus niger
Mycotoxin Type	Aflatoxin Group: B1, B2, G1, G3 Gliotoxin Derivative
Notes / Comments:	Composite Sample. Live mold growth evident. Results reported from redundancy samples[9]. As mold and mycotoxins detected, Validation Sampling will be required.

9   Sample analysis reported from redundancy sample material. Original results inconclusive from active mold growth sampled, as well as reported results inadvertently mixed with results from a non-client location.

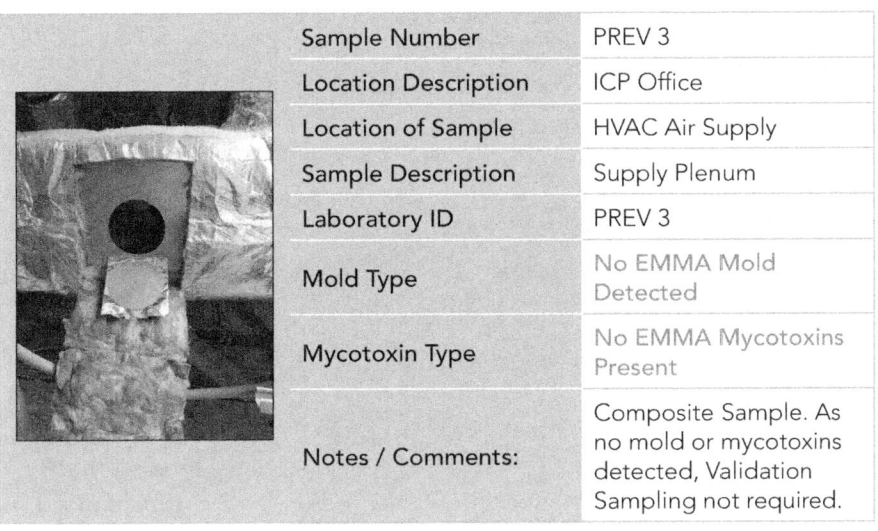

Sample Number	PREV 3
Location Description	ICP Office
Location of Sample	HVAC Air Supply
Sample Description	Supply Plenum
Laboratory ID	PREV 3
Mold Type	No EMMA Mold Detected
Mycotoxin Type	No EMMA Mycotoxins Present
Notes / Comments:	Composite Sample. As no mold or mycotoxins detected, Validation Sampling not required.

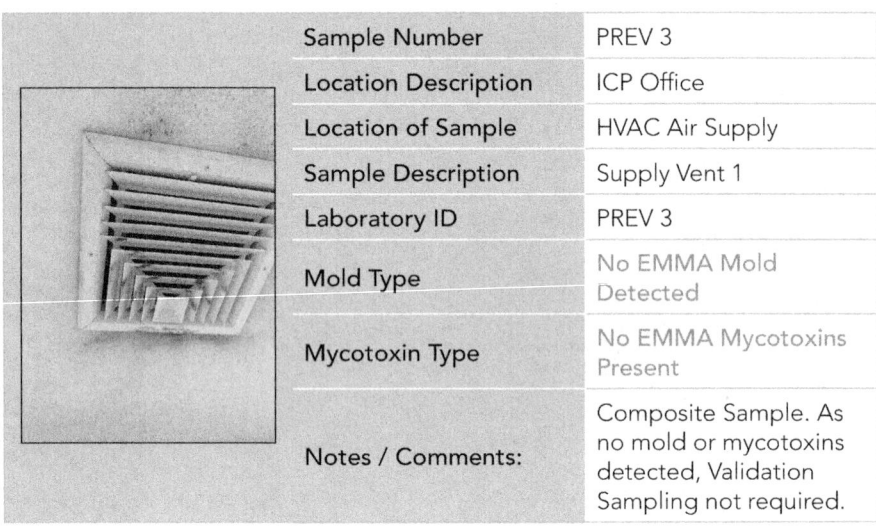

Sample Number	PREV 3
Location Description	ICP Office
Location of Sample	HVAC Air Supply
Sample Description	Supply Vent 1
Laboratory ID	PREV 3
Mold Type	No EMMA Mold Detected
Mycotoxin Type	No EMMA Mycotoxins Present
Notes / Comments:	Composite Sample. As no mold or mycotoxins detected, Validation Sampling not required.

Sample Number	PREV 3
Location Description	ICP Office
Location of Sample	HVAC Air Supply
Sample Description	Supply Vent 2
Laboratory ID	PREV 3
Mold Type	No EMMA Mold Detected
Mycotoxin Type	No EMMA Mycotoxins Present
Notes / Comments:	Composite Sample. As no mold or mycotoxins detected, Validation Sampling not required.

QMD PROTOCOL^{SM} TREATMENT

Risk Management: Decisions and Reasons

This document identifies the decisions, and reasons for the decisions, about the management of the risk(s) arising from any mold and mycotoxins identified and decontaminated in this report.

As mold and mycotoxins have been detected and located in this location, and as these may be associated with serious health concerns, this report's assessment of them are as follows:

- All areas identified as containing mycotoxins and the complete HVAC System be decontaminated via biohazard treatment; and
- The corresponding reductions of such recorded.

Control Measure Options

Health Risk Control Measures

Areas identified in this report as containing mold and mycotoxins are judged to constitute a potential health risk and have been decontaminated via biohazard treatment.

Decontamination Via Biohazard Treatment

Decontamination and biohazard treatment are not without hazards to the occupants of any home or workplace. If not strictly controlled, the decontamination process can result in increased airborne mold and mycotoxin counts in other areas than as identified.

A decontamination company's technical competence, experience, and integrity with any prescribed decontamination treatment for identified mold and mycotoxins, as well as the evaluation of their decontamination documentation and treatment protocols, are of prime importance.

As part of this treatment, please see the <u>Decontamination – Treatment</u> as itemized herein.

Decontamination Timetable
Identification Sampling: September 12th, 2020
Treatment: September 16th, 2020
Validation Sampling: September 16th, 2020

DECONTAMINATION – TREATMENT

Due to the evidence of live mold and the subsequent laboratory-confirmed presence of mycotoxins in the client's HVAC Air Returns and Air Handler Mechanism, it has been deemed necessary to firstly balance the office environment and deactivate any mycotoxins within this envelope via ionized H^2O^2, and secondly – and more specifically – treat the HVAC system using the same method.

Ionized Hydrogen Peroxide Overview
A low percentage of hydrogen peroxide (H_2O_2) is ionized into a solution of submicron particles – an exceptionally fine mist – that moves like a gas. The H_2O_2 molecules are turned into a Reactive Oxygen Species (ROS), mainly the hydroxyl radicals ($\cdot$OH) that are its killing/deactivation agents. $\cdot$OH are one of the most powerful oxidizing agents in nature – often referred to as 'nature's own disinfectant.'

During the process, the ionized H_2O_2 kills pathogens and fungal spores and deactivates viral cells and microorganisms on contact by destroying their proteins, carbohydrates, and lipids. This leads to the cellular disruption and deactivation, allowing for the quick decontamination of targeted areas, objects, and confined or restricted spaces.

Efficacy Process
The uniqueness of H_2O_2 when ionized effectively, takes an $\cdot$OH, previously known to only live for milliseconds, and effectively transfers them from "point A to point B," keeping the $\cdot$OH active and resulting in the destruction of pathogens, fungal spores, and microorganisms within seconds.

Ionized H_2O_2 allows for an effective kill in previously inaccessible and hard-to-reach areas, making it an ideal decontamination solution for contaminated HVAC systems.

Treatment Areas
Office Envelope and complete HVAC System.

The Process
When treating HVAC systems, the environment that the system draws from needs to first be charged so that, when the system is activated, it draws in the ionized mist in saturated draughts. Additionally, the environment needs to be charged so that when the mist is drawn into the HVAC system, all the internal circumferences of the ducts, plenums, and air handler are reached, thus avoiding the mist being just drawn down the middle of the system.

Dosage, Cycles & Injection Time
Due to the complex prerequisites required for the dosage, cycles, and injection times necessary to deactivate mycotoxins, a fully automated, remote-controlled, complete room decontamination fogging system will be used.

It is desired to have as dry an ionized mist as possible. The flow rate for fogging applicators will be set accordingly to achieve this.

The dosage and cycles of injection are calculated via, firstly, the cubic footage of the environment to be charged, and secondly, the cubic footage of the HVAC system to be treated.

Extensive testing was undertaken on an "open" HVAC System similar in size to the client's. Chemical peroxide test papers were adhered to the "open" air supply vents to determine the level and extent of the ionized solution's saturation within the system. As a result of testing, it has been determined that a multi-cycle, split mode injection would be required for treatment.

A multi-cycle, split mode injection of ionized H_2O_2 will be used for the treatment.

There will be a set injection time for each cycle. Initially turned off, the HVAC system will be activated after a specific time delay, partway through the injection phase of the cycle and allowed to remain on for the duration of the cycle. A dwell time will additionally be set between cycles. This is the "time between charges" that it takes the HVAC to draw all the visible mist from the room.

The above injection time will be repeated for the desired number of cycles.

Please Note
It is critical that accurate cubic footage, dosages, cycles, and injection times are calculated so an effective decontamination and biohazard treatment is achieved. Validation Sampling will show if this has been achieved or not.

Further Recommendations
It is further recommended:

- All impacted return filters be discarded and replaced.
- All contaminated insulation material within the air handler be discarded and replaced.
- As the floor drain in the secondary office is not being used, it be sealed as it is currently acting as an open water source.
- The office environment's humidity be reduced to approximately 50% relative humidity. This can either be done by running a thermostat with a humidistat or by running independent dehumidifiers.
- All ceilings painted with anti-fungal paint.

Treatment Images

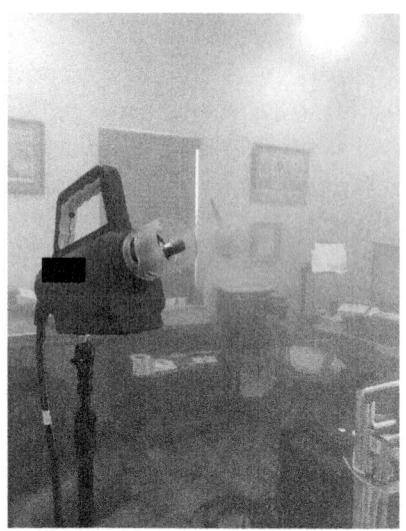

Main Office Deactivation –
Applicator 1

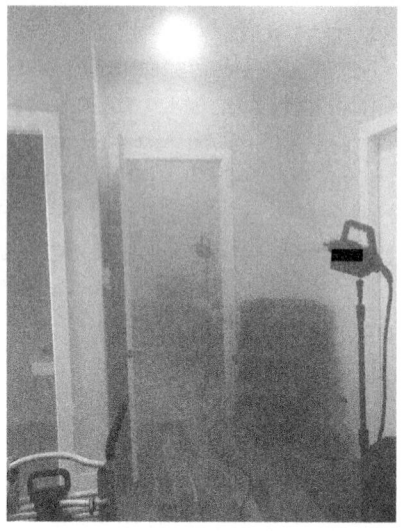

Main Office Deactivation –
Applicator 2

Secondary Office
Deactivation – Applicator 3

QMD PROTOCOL^{SM} VALIDATION –
MOLD & MYCOTOXIN REDUCTION

Location and Identification of Mold & Mycotoxin – Validation

Location Description	ICP Office	ICP Office
Location of Sample	HVAC Air Return	HVAC Air Handler
Sample Description	Composite: Floor Return Ducting, Ceiling Return Ducting Return Plenum	Composite: Cover Insulation, Core Drain Tray, Core Drain Tray Insulation
Sample Number	POSTV 4	POSTV 5
Laboratory ID	POSTV 4	POSTV 5
Type of Mold	Aspergillus niger (0.2451ng of DNA/mL)	No EMMA Mold Detected
Type of Mycotoxin	No EMMA Mycotoxins Present	No EMMA Mycotoxins Present

Site Comments – After Treatment & Validation Sampling

- This mycotoxin deactivation treatment and validation sampling has achieved its objective by giving a 100% reduction on all detected mycotoxins.
- While all mycotoxins were deactivated, a substantially reduced presence of mold was additionally quantified. As mycotoxins were not detected, the identified mold in Sample POSTV 4 may have also been deactivated and or rendered dormant. Although there was a total reduction of some mold and a substantial reduction (+98%) in Sample POSTV 4, the client has requested an additional decontamination treatment, targeting the HVAC Air Return, to remove all detected mold.
- Please see the Decontamination Treatment used and subsequent Validation Sampling to quantify the reduction below.

Reduction of Mold & Mycotoxins

From PREV 1 to POSTV 4 – ICP Office HVAC Air Return

Mold

ID Sample PREV 1 Type of Mold	Aspergillus fumigatus (0.4927ng of DNA/mL)	Aspergillus niger (13.3148ng of DNA/mL)	Chaetomium globosum (0.03ng of DNA/mL)
Validation Sample POSTV 4 Type of Mold	No EMMA Mold Detected	Aspergillus niger (0.2451ng of DNA/mL)	No EMMA Mold Detected
Reduction	100%	98.159%	100%

Mycotoxins

ID Sample PREV 1 Type of Mycotoxins	Ochratoxin A (12.62ppb)	Aflatoxin Group (1.727ppb)	Trichothecene Group (Macrocyclic) (0.042ppb)	Gliotoxin Derivative (0.868ppb)
Validation Sample POSTV 4 Type of Mycotoxins	No EMMA Mycotoxins Present	No EMMA Mycotoxins Present	No EMMA Mycotoxins Present	No EMMA Mycotoxins Present
Reduction	100%	100%	100%	100%

Reduction of Mold & Mycotoxins

From PREV 2 to POSTV 5 – ICP Office HVAC Air Handler

Mold

ID Sample PREV 2 Type of Mold	Aspergillus niger (3.7282ng of DNA/mL)
Validation Sample POSTV 5 Type of Mold	No EMMA Mold Detected
Reduction	100%

Mycotoxins

ID Sample PREV 2 Type of Mycotoxins	Aflatoxin Group (1.283ppb)	Gliotoxin Derivative (9.884ppb)
Validation Sample POSTV 5 Type of Mycotoxins	No EMMA Mycotoxins Present	No EMMA Mycotoxins Present
Reduction	100%	100%

Detailed Validation Sample Results

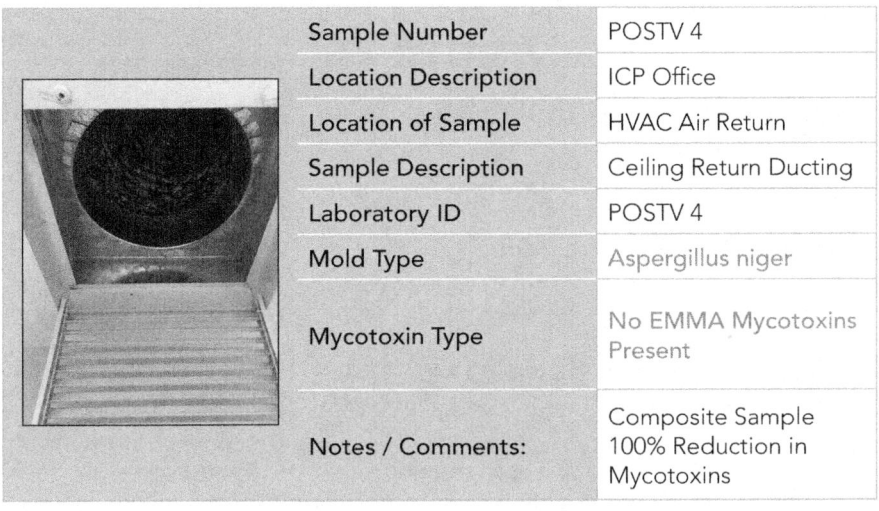

Sample Number	POSTV 4
Location Description	ICP Office
Location of Sample	HVAC Air Return
Sample Description	Floor Return Ducting
Laboratory ID	POSTV 4
Mold Type	Aspergillus niger
Mycotoxin Type	No EMMA Mycotoxins Present
Notes / Comments:	Composite Sample 100% Reduction in Mycotoxins

Sample Number	POSTV 4
Location Description	ICP Office
Location of Sample	HVAC Air Return
Sample Description	Ceiling Return Ducting
Laboratory ID	POSTV 4
Mold Type	Aspergillus niger
Mycotoxin Type	No EMMA Mycotoxins Present
Notes / Comments:	Composite Sample 100% Reduction in Mycotoxins

Sample Number	POSTV 4
Location Description	ICP Office
Location of Sample	HVAC Air Return
Sample Description	Return Plenum
Laboratory ID	POSTV 4
Mold Type	Aspergillus niger
Mycotoxin Type	No EMMA Mycotoxins Present
Notes / Comments:	Composite Sample 100% Reduction in Mycotoxins

Sample Number	POSTV 5
Location Description	ICP Office
Location of Sample	HVAC Air Handler
Sample Description	Cover Insulation
Laboratory ID	POSTV 5
Mold Type	No EMMA Mold Detected
Mycotoxin Type	No EMMA Mycotoxins Present
Notes / Comments:	Composite Sample 100% Reduction in Mycotoxins

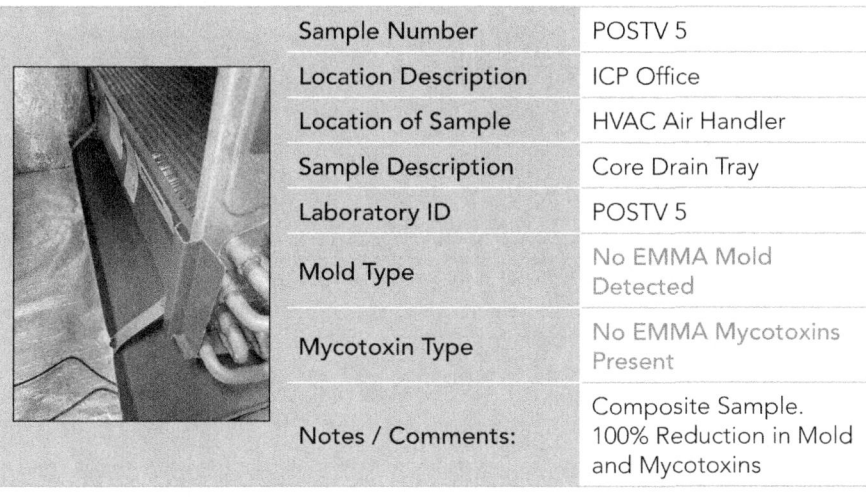

Sample Number	POSTV 5
Location Description	ICP Office
Location of Sample	HVAC Air Handler
Sample Description	Core Drain Tray
Laboratory ID	POSTV 5
Mold Type	No EMMA Mold Detected
Mycotoxin Type	No EMMA Mycotoxins Present
Notes / Comments:	Composite Sample. 100% Reduction in Mold and Mycotoxins

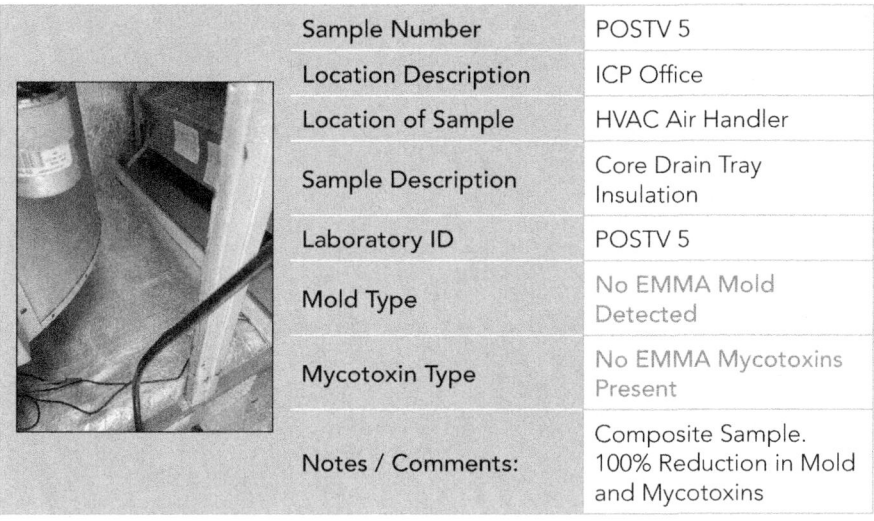

Sample Number	POSTV 5
Location Description	ICP Office
Location of Sample	HVAC Air Handler
Sample Description	Core Drain Tray Insulation
Laboratory ID	POSTV 5
Mold Type	No EMMA Mold Detected
Mycotoxin Type	No EMMA Mycotoxins Present
Notes / Comments:	Composite Sample. 100% Reduction in Mold and Mycotoxins

ADDITIONAL DECONTAMINATION – TREATMENT PROTOCOL

Due to Validation laboratory analysis detection of Aspergillus niger in Sample POSTV 4, and although no mycotoxins were detected in this sample, the client has requested an additional decontamination treatment be undertaken on the HVAC Air Return housings. Please see the report section titled QMD Protocol[SM] Treatment for the technology, equipment, and efficacy process of the protocol that will be used.

Additional Decontamination Timetable
Additional Decontamination Treatment Protocol: October 7th, 2020
Validation Sampling: October 7th, 2020

Decontamination, Biohazard Treatment, and Forensic Clean
The treatment will undertake multiple Biohazard applications, HEPA Vacuuming, and a Forensic Clean.

The ionized H_2O_2 application will be undertaken by a handheld surface treatment. This will give the technician the freedom to easily manipulate the directional flow of the ionized mist and specific targeting of all surfaces of the air return housings.

The flow rate for all mist applicators used for this protocol will be set to produce a semi-dry submicron mist. The square foot surface area to be treated will be calculated so standard operating application procedures of the ionized solution will be applied to correctly treat the area.

The first ionized H_2O_2 application will be applied as a pre-treatment load reduction to the client's HVAC Air Return housings and ducting to assist in balancing the areas.

After the ambient environment has been reduced below 1ppm of H_2O_2, all return filter housings will be HEPA vacuumed to remove any debris left after the removal of the impacted filters prior to the original ionized H_2O_2 decontamination treatment.

After HEPA vacuuming is complete, a full Forensic Clean will be undertaken, consisting of a dry microfiber wipe-down, and a full microbial fungicide disinfectant treatment of the HVAC Air Return housings.

After the microfiber and microbial fungicide treatment are complete, a final ionized H_2O_2 application will be undertaken.

Please Note
It is critical that the correct square footage, dosages, and application rates are calculated and used so an effective decontamination and biohazard treatment is achieved. Validation Sampling will show if this has been achieved or not.

Additional Decontamination Images

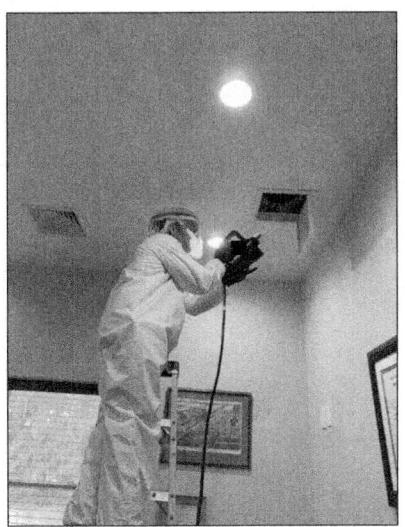 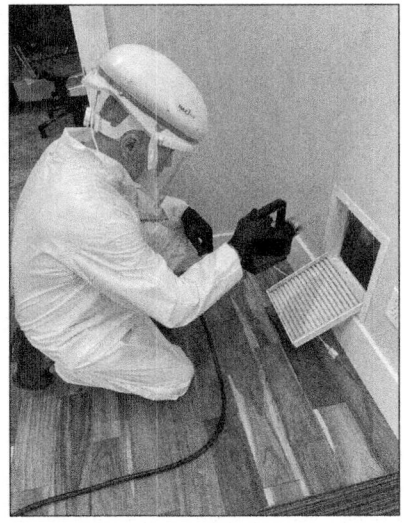

Ceiling Return Housing
Decontamination – via
ionized H_2O_2 Surface
Treatment

Floor Return Housing
Decontamination – via
ionized H_2O_2 Surface
Treatment

ADDITIONAL MOLD REDUCTION

Location and Identification of Mold – Validation

Location Description	ICP Office
Location of Sample	HVAC Air Return
Sample Description	Composite: Floor Return Ducting, Ceiling Return Ducting Return Plenum
Sample Number	POSTV +1
Laboratory ID	POSTV +1
Type of Mold	No EMMA Mold Detected
Type of Mycotoxin	No EMMA Mycotoxins Present

Site Comments – After Additional Decontamination & Validation Sampling
- This mycotoxin sampling, decontamination and deactivation treatment has achieved its objective by deactivating all mold and mycotoxins detected.
- See below: the 100% reduction of mold between samples POSTV 4 and validating sample POSTV +1.

Reduction of Mold

From POSTV 4 to POSTV +1 – ICP Office HVAC Air Handler

Mold

Validation Sample POSTV 4 Type of Mold	Aspergillus niger (0.2451ng of DNA/mL)
Post-Validation Sample POSTV +1 Type of Mold	No EMMA Mold Detected
Reduction	100%

Mycotoxins

Validation Sample POSTV 4 Type of Mycotoxins	No EMMA Mycotoxins Present
Post-Validation Sample POSTV +1 Type of Mycotoxins	No EMMA Mycotoxins Present
Reduction	N/A

Detailed Validation Sample Results

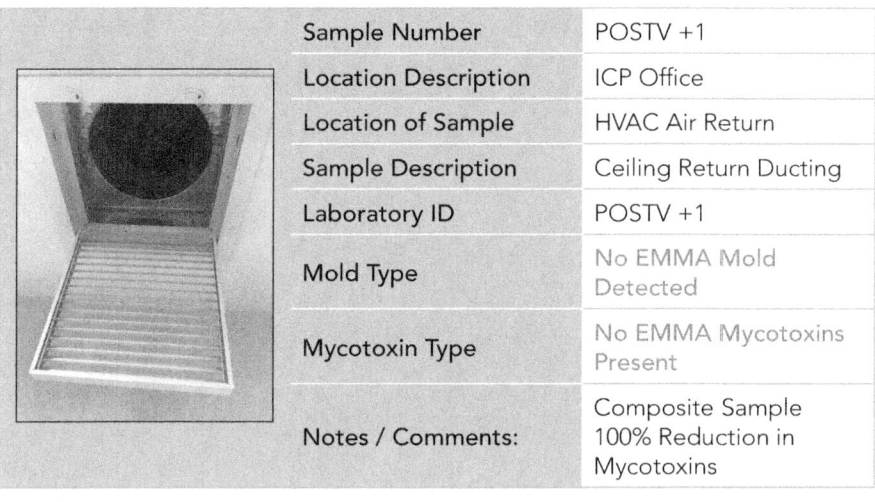

Sample Number	POSTV +1
Location Description	ICP Office
Location of Sample	HVAC Air Return
Sample Description	Floor Return Ducting
Laboratory ID	POSTV +1
Mold Type	No EMMA Mold Detected
Mycotoxin Type	No EMMA Mycotoxins Present
Notes / Comments:	Composite Sample 100% Reduction in Mycotoxins

Sample Number	POSTV +1
Location Description	ICP Office
Location of Sample	HVAC Air Return
Sample Description	Ceiling Return Ducting
Laboratory ID	POSTV +1
Mold Type	No EMMA Mold Detected
Mycotoxin Type	No EMMA Mycotoxins Present
Notes / Comments:	Composite Sample 100% Reduction in Mycotoxins

	Sample Number	POSTV +1
	Location Description	ICP Office
	Location of Sample	HVAC Air Return
	Sample Description	Return Plenum
	Laboratory ID	POSTV +1
	Mold Type	No EMMA Mold Detected
	Mycotoxin Type	No EMMA Mycotoxins Present
	Notes / Comments:	Composite Sample 100% Reduction in Mold and Mycotoxins

FLOOR PLAN & SAMPLE LOCATIONS OVERVIEW

2710 Brantley Blvd

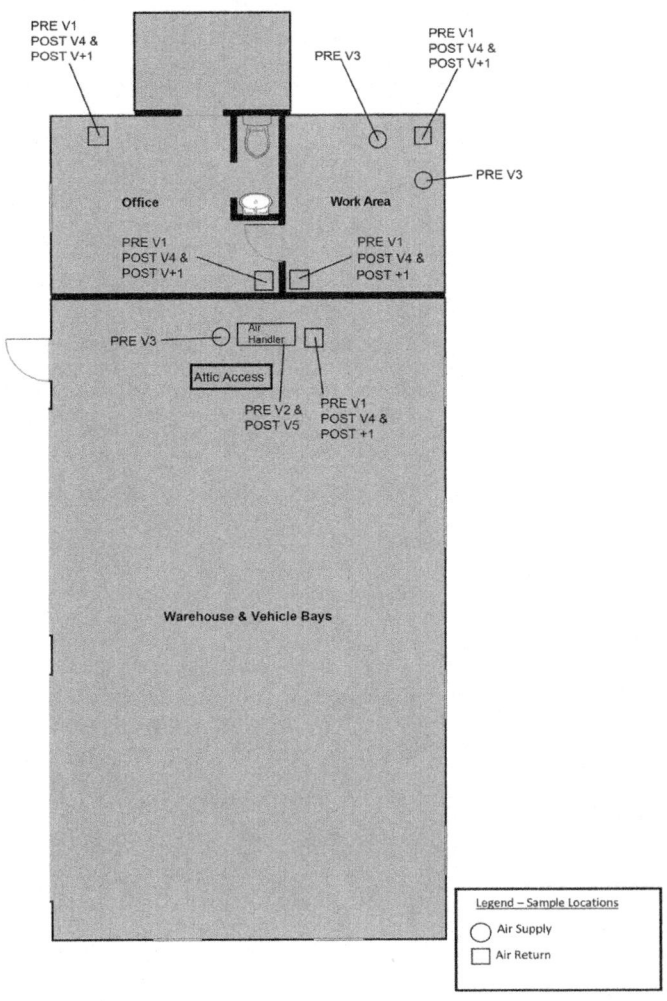

CONCLUSIONS

The documentation and Validation Sampling laboratory analysis of this report are sufficient to establish:
- There existed a prior mold and mycotoxin contamination in the client's office HVAC system.
- That deactivation via an ionized H_2O_2 decontamination treatment was undertaken.

Additionally, Validation Sampling showed:
1. Efficacy and effectiveness of all decontamination treatment produced a total reduction in mycotoxin contamination.
2. The QMD Protocol[SM] achieved its objective, and all detected mycotoxins were deactivated.
3. All originally detected mold was subsequently removed.

The independent laboratory data supplied as part of this report is sufficient to confirm the authenticity of the results reported and documented.

The photos provided throughout the report demonstrate comprehensive identification and validation sample locations, numerous site images, and the documented decontamination treatment in process.

It is the conclusion of the author that HEPA vacuuming of the return filter housings of debris left after the removal of the impacted/contaminated filters prior to undertaking the initial decontamination treatment would have eliminated the need for additional decontamination and validation sampling.

The desired result of deactivation and subsequent reduction of the mycotoxin previously detected by Identification Sampling in the client's HVAC system has been successfully achieved, and the environment restored back to balance.

Want more information, or would like to contact the author?
Email info@biorisk.us, call (877) BioRisk,
or visit www.biorisk.us/mycotoxin_deactivation

LIMITATIONS AND DISCLAIMERS

BioRisk Reports are conducted in a conscientious and professional manner. The nature of the task and the likely disproportion between any damage or loss which might arise from the work or reports prepared, and the cost of our services, is such that BioRisk cannot guarantee that all mold and mycotoxins have been identified and/or addressed.

Due to the possibility of renovations and additions to the buildings over time, mold and mycotoxins may have formed/grown behind new walls, ceilings, and flooring where such areas were inaccessible during the inspection. If any areas reasonably suspected of containing mold and mycotoxins are found during further renovation and/or demolition of the buildings, the material should be sent for identification, and expert advice sought.

Thus, while we carry out the work to the best of our ability, we totally exclude any loss or damages which may arise from services we have provided to the client and/or their associated parties.

Associated Parties
All work conducted and reports produced by BioRisk are prepared for a particular client's objective and are based on a specific scope of work, conditions, and limitations, as agreed upon between BioRisk and the client. Information and/or report(s) prepared by BioRisk may therefore not be suitable for any use other than the intended objective.

This report, including any recommendations, is based upon site observation. It is possible that unknown and or hidden conditions may exist which would influence this report, its conclusions, and recommendations. As such, BioRisk and associated parties to this report expressly disclaims any liability to any parties, client or otherwise, that may rely on this report and its conclusions and recommendations.

BioRisk, its employees, and any associated parties have exercised the degree of skill and care expected and customarily accepted by good architectural practices and procedures, as well as environmental practices and procedures. As such, no other warranties, expressed or implied, are made with respect to their performance unless agreed by all parties in writing. Use of any portions of this report out of context, not fully assembled with exhibits, and without original signature shall not be the responsibility of BioRisk, its employees, or any associated parties.

Disclaimers

Information reported herein is based on the interpretation of data collected from third parties and has been accepted in good faith as being accurate and valid.

This report is for the exclusive use of the client. No warranties or guarantees are expressed or should be inferred by any third parties. BioRisk and associated parties disclaim any responsibility to the client and others in respect of any matters specifically outside the agreed scope of work.

INTELLECTUAL PROPERTY & TRADE SECRET RIGHTS

Except for rights expressly granted via original signature, nothing in this report will function to transfer any of the Author's and BioRisk's Intellectual Property and Trade Secret rights to the client, any associated parties, or recipients of this report. BioRisk retains exclusive interest in and ownership of its Intellectual Property and Trade Secret(s) developed before, during, or outside the scope of this report, case study and or publication (report).

Trade Secret Protection

The Author and BioRisk has taken all reasonable measures and precautions to protect and maintain the confidentiality and value of all its Trade Secrets included in by or outside of its Intellectual Property. The Author and BioRisk has not disclosed any Trade Secrets in which it has (or purports to have) any right, title, or interest (or any tangible embodiment thereof) to any Person without having such Person execute a valid, binding, enforceable written agreement regarding the non-disclosure and limitations on use thereof. Being a recipient of this report in no way invalidates the Author's and BioRisk's Trade Secret protection rights.

Work Product

The Work Product, and all documentation, information, systems, and other results developed in connection with the Work Product, will, to the extent permitted by law, be a "work made for hire" within the definition of the Copyright Act (17 U.S.C. 101) and will remain the Author's and BioRisk's exclusive property. If, and to the extent that any, Work Product is not deemed to be a work made for hire within the definition of the Copyright Act, at the completion and acceptance of this report, the client shall promptly assign to the Author and or BioRisk all its right, title, and interest in and to the Work Product, including any Intellectual Property rights.

Service

The Author and BioRisk will retain all interest in and to the Service, including all documentation, modifications, improvements, upgrades, derivatives, and all other Intellectual Property and Trade Secret rights in connection with the Service, including but not limited to the Author's and or BioRisk's name, logos, trademarks and service marks reproduced through the Service.

Use and Acceptance

If this report, Work Product, or Service is not accepted, or a contractual agreement for such is disputed, by the Client and or associated parties, this Report must be returned to BioRisk. Any use of, reference to, or discussion of, without the Author and BioRisk's permission and or compensation of such, is consider a violation of the Author's and BioRisk's Intellectual Property and Trade Secrets rights under Copyright Law 17 U.S.C. 102(a), 106, and Trade Secret Law 18 U.S.C. 1831, 1832, and will be pursued as such.

APPENDICES

References

Mycotoxins: In the publication _Mycotoxins: Children's Health and the Environment_, the World Health Organization (WHO) defines mycotoxins as "[n]atural products produced by fungi that evoke a toxic response when introduced in low concentrations to higher vertebrates by a natural route."

Note: The WHO does not specifically define "low concentration," humans are higher vertebrates, and inhalation is a natural route.

Aflatoxins: On their website, the National Cancer Institute (NIH) notes that "[e]xposure to aflatoxins is associated with an increased risk of liver cancer."

Ochratoxins: The U.S. Department of Health and Human Services' 14th Report on Carcinogens (RoC) states aflatoxins are "[k]nown to be a Human Carcinogen" and Ochratoxin A as "[r]easonably anticipated to be Human Carcinogen."

Gliotoxin: The NIH states that gliotoxin is an immunosuppressive mycotoxin long suspected to be a potential virulence factor of _Aspergillus fumigatus_.

Trichothecenes: Centers for Disease Control and Prevention's (CDC) Case Definition of Trichothecene Mycotoxin states: "The trichothecene mycotoxins are a group of toxins produced by multiple genera of fungi."

They later state: "Systemic symptoms can develop with all routes of exposure (especially inhalation) and might include weakness, ataxia, hypotension, coagulopathy and death."

Mycophenolic Acid: As verified by the NIH, exposure to mycophenolic acid during pregnancy is associated with increased risks of pregnancy loss and congenital malformations. Females of reproductive potential must be counseled regarding pregnancy prevention and planning. Increased risk of development of lymphoma and other malignancies, particularly of the skin, due to immunosuppression and increased susceptibility to bacterial, viral, fungal, and protozoal infections, including opportunistic infections, are also risks.

Sterigmatocystin: Sterigmatocystin is carcinogenic to mice (causing pulmonary adenocarcinomas) and rats (causing hepatocellular carcinomas at milligram doses of sterigmatocystin per animal per day for 1 year) following oral administration. It is classified as an International Agency for Research on Cancer class 2B carcinogen (i.e., possibly carcinogenic to humans), as detailed in the Applied and Environmental Microbiology journal.

Chaetoglobosins: Chaetomium globosum, the most common species within the genus, produces chaetoglobosins A and C when cultured on building material. The NIH notes that relatively low levels of these compounds have been shown to be lethal to various tissue culture cell lines.

EMMA Environmental Test Specifications

Environmental Test Specifications

EMMA DNA Panel

Background:

Mold in a house isn't just a problem for people with allergies or asthma. Water leaks in homes provide an ideal environment for mold growth. There are many mold species, black mold or Stachybotrys Chatarum is more dangerous than the most common indoor molds such as Aspergillus and Penicillium. Mold exposure can cause immunosuppression and upper respiratory infections. Long term exposure to mold can cause acute to chronic illnesses.

Test Description:

The EMMA test is a quantitative PCR (qPCR) procedure for the detection of ten pathogenic fungal species in environmental dust specimens. EMMA includes six assays that were designed and used by the EPA and four assays that were previously developed by RTL. The qPCR method used in these assays utilizes the hybridization of a species-specific probe to a complimentary DNA strand to amplify and detect fungal DNA. The data generated for each specimen is plotted against a standard curve to calculate the amount of DNA present in the specimen (nanograms of DNA per milliliter of dust in PBS buffer). A process control (Geotrichum) is included to verify that the DNA extraction procedure was successful, and PCR positive controls are run with each amplification.

References:

- The Biocontaminants and Complexity of Damp Indoor Spaces: More than What Meets the Eyes (Authors: Thrasher JD and Crawley S). *Toxicology and Industrial Health.* 2009.
- Mycotoxin Detection in Human Samples from Patients Exposed to Environmental Molds. Hooper, D.G., Bolton, V.E., Guilford, F.T and D.C. Straus. Int. J. Mol. Sci. 2009, 10, 1465-1475
- Indoor Environmental quality – Dampness and mold in the buildings https://www.cdc.gov/niosh/topics/indoorenv/mold.html.

4100 Fairway Ct. - Suite 600
Carrollton, TX 75010
www.realtimelab.com
972-492-0419

Assay Method: Quantitative PCR (qPCR)

Mold DNA Targets:

Aspergillus fumigatus	*Aspergillus ochraceus*	*Candida auris*
Aspergillus flavus	*Aspergillus terreus*	*Chaetomium globosum*
Aspergillus niger	*Aspergillus versicolor*	*Fusarium solani*
		Stachybotrys chartarum

PCR Amplification Efficiency

Amplification efficiency was evaluated by obtaining concentrated DNA from an independent vendor for all assays and serially diluted ten-fold to produce dilutions of 1/10, 1/100, 1/1000, and 1/10000. The dilutions for each assay were amplified in triplicate to obtain amplification efficiency. All assays demonstrated a high amplification efficiency. Amplification efficiency was calculated using the following equation:

$$\text{Efficiency} = -1 + 10^{\wedge(-1/\text{slope})} \div 100\%$$

Assay	A. flavus	A. fumigatus	A. niger	A. ochraceus	A. terreus
Efficiency	77.7%	86.0%	107.6%	76.4%	96.4%

Assay	A. versicolor	C. auris	C. globosum	F. solani	S. chartarum
Efficiency	67.3%	93.7%	105.1%	97.2%	98.7%

Precision/Reproducibility

Assay precision was determined by testing twenty replicates of positive controls over several days and between multiple technologists. Inter-run cycle threshold values display high precision with less than 5% CV in all assays.

Linearity

All qPCR assays are highly linear ($R^2 > 0.97$) over several orders of magnitude.

Limit of Detection (LOD)

Assay LOD was determined for all assays by obtaining concentrated DNA and diluting the DNA down to 1000 nanograms per mL of dust in PBS buffer. DNA was further serially diluted ten-fold down to 0.001 ng/mL. LOD samples were processed in triplicate. The amount of DNA (ng/mL) detected in >95% of replicates is presented below:

Assay	A. flavus	A. fumigatus	A. niger	A. ochraceus	A. terreus
>95% Detection	0.657	0.012	0.015	1.810	1.393

Assay	A. versicolor	C. auris	C. globosum	F. solani	S. chartarum
>95% Detection	2.625	0.065	0.038	0.314	0.292

Specificity

Assay Specificity was determined by obtaining purified DNA and processing with each assay. All prepared samples were run with all assays in triplicate. All assays show 100% specificity for their intended target.

Accreditation

RealTime Laboratories, Inc. is a CAP (#7210193) and CLIA (#45D1051736) accredited testing laboratory.

1

Environmental Test Specifications

Background

Mycotoxins are small molecular weight toxic molecules produced by various species of mold, some of which inhabit water damaged homes or buildings. Many clinical symptoms and disease states have been associated with human exposure to mycotoxins. The RealTime Lab Mycotoxin Panel detects 15 different mycotoxins, as follows: Trichothecenes (Satratoxin G and H, Isosatratoxin F, Roridin A, E, H, and L-2, and Verrucarin A and J.), Ochratoxins (Ochratoxin A), Aflatoxins (Aflatoxin B1, B2, G1, and G2), and Gliotoxin (bis (methyl) gliotoxin)

Mycotoxins are particularly important because they are known to be highly toxic and are produced by species such as Stachybotrys ("black mold"), shown to be present in mold contaminated buildings.

Testing is done using competitive ELISA, a very sensitive and specific method for detection using antibodies prepared against the Mycotoxins.

References:
- Mycotoxin Detection in Human Samples from Patients Exposed to Environmental Molds. Hooper, D.G., Bolton, V.E., Guilford, F.T. and D.C. Straus. Int. J. Mol. Sci. 2009, 10, 1465-1475
- Chronic Illness Associated with Mold and Mycotoxins: Is Naso-Sinus Fungal Biofilm the Culprit? Brewer, J.H., Thrasher, J.D., and D. Hooper. Toxins. 2014. Jan; 6(1):66-80
- Intranasal Antifungal Therapy in Patients with Chronic Illness Associated with Mold and Mycotoxins: An Observational Analysis. Brewer, J.H., Hooper, D., and S. Muralidhar. Global J. of Med. Res. 2015. 15(1). 29-33
- Trichothecenes: From Simple to Complex Mycotoxins. McCormick, S.P., Stanley, A.M., Stover, N.A., and N.J. Alexander. Toxins. 2011. 3, 802-814
- Enzyme Immunoassay for the Macrocyclic Tricothecene Roridin A: Production, Properties and use of Rabbit Antibodies. Martlbauer, E.,

4100 Fairway Ct. - Suite 600
Carrollton, TX 75010
www.realtimelab.com
972-492-0419

EMMA Mycotoxin Panel

Assay Method: ELISA

Accuracy
Assay accuracy was evaluated by obtaining concentrated mycotoxin from an independent vendor, and the RTL Mycotoxin Panel was used to measure the concentration of each mycotoxin present in a range of dilutions specific for each assay. Measurements must be accurate within 20% of the expected value for samples measured in the assay specific ranges.

Assays	Tested Concentration Range (PPB)	Percent Error
Trichothecene	0.01 to 1.0	≤ 20.0
Ochratoxin	0.5 to 10.0	≤ 20.0
Aflatoxin	1.0 to 8.0	≤ 20.0
Gliotoxin	0.3 to 10.0	≤ 20.0

Precision/Reproducibility
Assay precision was determined by spiking a negative urine sample with a known amount of mycotoxin and testing ten replicates for each assay. Measurements must have a Coefficient of Variation (CV) of ≤ 20%.

Assays	Tested Concentration Range (PPB)	Coefficient of Variation (%CV)
Trichothecene	0.01 to 1.0	≤ 20.0
Ochratoxin	0.5 to 10.0	≤ 20.0
Aflatoxin	1.0 to 8.0	≤ 20.0
Gliotoxin	0.3 to 10.0	≤ 20.0

Linearity
The RTL Mycotoxin Assays are highly linear ($R^2 > 0.95$) over several orders of magnitude. The reportable ranges for the assays are as follows:

Assays	Present if ≥	Equivocal if between
Trichothecene	0.03 ppb	0.02-0.03 ppb
Ochratoxin	2.0 ppb	1.8-2.0 ppb
Aflatoxin	1.0 ppb	0.8-1.0 ppb
Gliotoxin	1.0 ppb	0.5-1.0 ppb

Sensitivity
The analytical limit of detection for the mycotoxin assays are as follows: Trichothecene is 0.01 ppb, Ochratoxin is 0.5 ppb, Aflatoxin is 1.0 ppb, and Gliotoxin is 0.3 ppb.

Specificity
The RTL Mycotoxin Panel is specific for the detection and measurement of 15 specific mycotoxins. Each assay is specific for the mycotoxins specified and do not cross react with any other mycotoxins in the same sample thus not yielding a false positive result.

Accreditation
RealTime Laboratories, Inc. is a CAP (#7210193) and CLIA (#45D1051736) accredited testing laboratory.

EMMA Fungal Load Explanation

Environmental Mold and Mycotoxin Assay (EMMA)

Fungal Load Explanation

"*Fungal load*" is a term that RealTime Laboratories will use when discussing fungal infestation and possible infection in the patient and his/her environment.

Without treatment in the body, the fungus can and usually does replicate (makes copies of itself) which causes the amount of fungus in the body to increase unchecked. This same thing happens in an environment like a house or a building. The fungus that is growing can and usually does make mycotoxins (mold toxins) which can prove detrimental to the body in many instances.
In the past, environmental testing laboratories have used numbers of spores/mg of dust or numbers of spores/ml of solution (reference ERMI and qPCR). The results are reported as the number of spores in the submitted specimen. The results can be misleading because of the difficulty in determining the number of spores in a colony of a filamentous (branching) fungus like *Aspergillus sp. or Penicillium sp.* Many of the more harmful fungal elements are filamentous. Thus, an actual spore count is difficult to conduct and is often misleading to the physician, building inspector, and patient. The difficulty of interpretation has led to a mistrust of laboratory results or interpretation and has left the industry in a quandary of deciding levels of fungal contamination that are dangerous to the population.

RealTime Laboratories has developed a more accurate method of reporting. Dust is homogenized in buffered saline prior to lab testing. The numbers of spores in a culture of fluid (spores)/ml of fluid are then documented and compared to the actual amount of DNA in ng/ml in the same fluid. RealTime's EMMA results will now read DNA ng/ml of fluid.

This is simply a measurement of how much fungal DNA is present in a sample. This will be more beneficial to the patient, physician, and the environmental inspector when evaluating fungal infestation in the home, etc., as well as evaluating the efficacy of the treatment of the environment.

The aim of mycotoxin and fungal treatment is an undetectable mycotoxin level in your body and the same undetectable results in your home. Your mycotoxin/fungal load should have fallen to undetectable levels within three to six months of starting treatment. If this doesn't happen, your doctor will talk to you about possible reasons for this and discuss what to do next. Once you have an undetectable fungal load in one or both areas, you should have your fungal load monitored every three to four months and your home measured once a year. If you have had an undetectable fungal load for some time and are doing well on treatment, your doctor may offer you the option to have your fungal load measured every year for two more years, both in your house and body.

P 972.492.0419
F 972.243.7759

4100 Fairway Court, Suite 600
Carrollton, TX 75010

CAP #7210193 CLIA #: 45D1051736

All fungal load tests have a cut-off point below which they cannot reliably detect fungi. This is called the limit of detection (LOD). If your fungal load or your house's fungal load is below LOD such readings can be considered undetectable. But just because the level of mycotoxin and/or fungus is too low to be measured doesn't mean that fungus and mycotoxins have disappeared completely from your body or your home. It might still be present but in amounts too low to be measured. Thus, the patient and environmental inspectors must be vigilant in their pursuit of ridding their body and the environment they live in of the molds and mycotoxins.

Readings of RTL are listed here with the correlation of 1 ng of DNA/ml to spores/ml. Dust is placed in 1 ml of Phosphate Buffered Saline and extracted for DNA. 200 ul of sample is used to test for DNA. Final conversions are reported in 1000 ul or 1 ml of solution.

Species	1 ng of detected DNA per mL of sample is an average equivalent to the following number of spores per mL:
Aspergillus flavus	1.70
Aspergillus fumigatus	11.10
Aspergillus niger	0.53
Aspergillus ochraceus	7.43
Aspergillus terreus	0.48
Aspergillus versicolor	2.26
Chaetomium globosum	2.92
Fusarium species	9.80
Stachybotrys chartarum	1.50

Species	1 ng of detected DNA per mL of sample is approximately equivalent to the following number of colonies forming units (CFU) per mL:
Candida auris **	0.0023[#]

** Candida auris is an emerging multidrug-resistant yeast causing invasive health care associated infection with high mortality (Chowdhary, et al, 2017, PLoS Pathog 13:e1006290). One colony of a Candida auris is removed from Sabouraud Dextrose Agar, lysed and DNA is analyzed and reported in ng/ml. Results obtained are reported in ng of DNA/ml of sample.
#. 1 CFU/ml of Candida auris is equivalent to 435ng/ml.

 972.492.0419
972.243.7759

4100 Fairway Court, Suite 600
Carrollton, TX 75010

www.RealTimeLab.com

CAP #7210193 CLIA #: 45D1051736

EMMA Mold Test Panel

REALTIME
LABORATORIES INC.
Cutting edge. Breaking barriers.

BRIEF EXPLANATION GUIDE ON ENVIRONMENTAL MOLD PANEL TESTING

MOLD	MYCOTOXIN PRODUCED	POTENTIAL HEALTH ISSUES
Aspergillus fumigatus	Gliotoxin, Aflatoxin	A. fumigatus is frequently found in homes and buildings. It is considered to be an opportunistic pathogen, meaning it rarely infects healthy individuals, but is the leading cause of invasive aspergillosis (IA) in immunocompromised individuals such as cancer, HIV or transplant patients.
Aspergillus flavus	Gliotoxin, Aflatoxin	A. flavus is the second leading cause of invasive aspergillosis in immunocompromised patients. Particularly common clinical syndromes associated with A. flavus include: chronic granulomatous sinusitis, keratitis, cutaneous aspergillosis, wound infections and osteomyelitis following trauma and inoculation. Can cause liver cancer in humans
Aspergillus terreus	Gliotoxin, Citirin	Inhalation of fungal spores, which travel down along the respiratory tract, cause the typical respiratory infection.
Aspergillus versicolor	Sterigmatocystin	A. versicolor is one of the most frequently found molds in water-damaged buildings. A. versicolor is known to produce a mycotoxin called sterigmatocystin a potentially carcinogenic and hepatotoxic mycotoxin. It is primarily toxic to the liver and kidneys.
Aspergillus ochraceus	Ochratoxin	Ochratoxin has been demonstrated to be Nephrotoxic, Hepatotoxic, and Carcinogenic and is a potent teratogen and immune-suppressant. It has also been associated with urinary tract infections and bladder cancers.
Aspergillus niger	Ochratoxin, Gliotoxin	A. niger produces gliotoxin, which has been identified in the sera of humans and mice with aspergillosis. Causes immunosuppression in patients.
Stachybotrys chartarum	Macrocyclic Trichothecenes	S. chartarum, commonly known as black mold, is highly toxic to humans. Nausea, vomiting, diarrhea, burning erythema, ataxia, chills, fever, hypotension, hair loss and confusion are symptoms in individuals living or working inside Stachybotrys infested homes and buildings.
Chaetomium globosum	Chaetoglobosins	C. globosum is a common indoor fungal contaminant of water damaged homes or buildings. Like Stachybotrys, C. globosum spores are relatively large and due to their mode of release are not as easily airborne as are some other molds.
Fusarium species	Fumonosins; Zearalanone	Fusarium can cause superficial infections such as keratitis or onychomycosis in healthy individuals and disseminated infections in immunocompromised patients.
Candida auris	Unknown	C. auris can be found in healthcare facilities and can be spread through contact with infected patients and equipment's. C.auris can cause blood stream infections, wound infections and ear infections.

For any further question on the test report please schedule a consult with our medical staff at www.realtimelab.com

972.492.0419
972.243.7759

4100 Fairway Court, Suite 600
Carrollton, TX 75010

CAP #7210193 CLIA #: 45D1051738

www.RealTimeLab.com

EMMA Mycotoxin Test Panel

BRIEF EXPLANATION ON MYCOTOXIN PANEL

For any further question on the test report please schedule a consult with our medical staff at www.realtimelab.com

	Mycotoxin	Cellular activity of Mycotoxin	Symptoms/Others	Association with a "Disease State"
		AFLATOXIN FAMILY—Organism: *Aspergillus flavus*, *Aspergillus oryzae*, *Aspergillus fumigatus*, *Aspergillus parasiticus*		
		Aflatoxins have been linked to liver cancer, hepatitis, cirrhosis, and other health issues		
1	B1	Binds DNA and proteins	Shortness of breath, weight loss, most potent and highly carcinogenic.	Primarily attacks the liver, other organs include kidneys and lungs.
2	B2	Inhibits DNA, RNA, and protein metabolism	Enters the body through the lungs, mucous membranes (nose and mouth), or the skin.	Affects the liver and kidneys. Less potent than B1. The order of toxicity is B1 greater than G1, greater than G2, greater than B2.
3	G1	Adversely affects the immune system	A. flavus second leading cause of invasive aspergillosis in immunocompromised patients.	Cancer, chronic hepatitis, and jaundice. Reye's Syndrome.
4	G2	Immunosuppression.	Mitochondrial damage. Aflatoxicosis in Humans and Animals.	Hepatitis, malnutrition, lung cancer.
		OCHRATOXIN A —Organisms: *Aspergillus ochraceus*, *Aspergillus niger*, and *Penicillium* species		
5	Ochratoxin A	Interferes with cellular physiology, inhibits mitochondrial ATP production, and stimulates lipid peroxidation.	A potent teratogen and immune— suppressant. 30—day ½ life in blood, indefinite existence intra—cellular.	Kidney disease, cancer, infection of bladder, Nephrotoxic, Hepatotoxic.
		TRICHOTHECENE FAMILY (MACROCYCLIC) —Group D Organism: *Stachybotrys chartarum*		
6	Satratoxin G	DNA, RNA and protein synthesis, intracellular	Bleeding disorders, central nervous and peripheral nerve disorders. Most potent inhibitors of protein synthesis.	Wide range of GI problems, skin inflammation, vomiting and damage to blood producing cells. Highly toxicosis.
7	Satratoxin H	Inhibits protein synthesis	Found in damp or water— damaged environment.	Vision problems, GI problems, breathing issues.
8	Isosatratoxin F	Immunosuppression	Causes of health problems due to poor air quality.	Contributor to "sick building syndrome"
9	Roridin A	Nasal inflammation, excess mucus secretion, and damage to the olfactory system	Acute and chronic respiratory tract health problems.	Acute and chronic lung and nasal problems.
10	Roridin E	Disrupt the synthesis of DNA, RNA, and protein	Roridin E grows in moist indoor environments.	Can impact every cell in the body
11	Roridin H	Inhibits protein synthesis	Grows well on many building materials subject to damp conditions.	Lymphoid necrosis and dysregulation of IgA production.
12	Roridin L-2	Immunosuppression	Grows on wood—fiber, boards, ceiling tiles, water—damaged gypsum board, and HVAC ducts.	Easily airborne and inhaled by occupants of an infected building.
13	Verrucarin A	Immunosupression, nausea, vomiting, weight loss	Found mostly in damp environments.	One of the most toxic trichothecenes.
14	Verrucarin J	Can easily cross cell membranes	Absorbed through the mouth and the skin.	Small enough to be airborne and easily inhaled.
		GLIOTOXIN DERIVATIVE—Organisms: *Aspergillus fumigatus*, *Aspergillus terreus*, *Aspergillus niger*, *Aspergillus flavus*		
15	Gliotoxin	Attacks intracellular function in immune system	Lung disorders, brain dysfunction, bone marrow dysfunction.	Immune dysfunction disorders. Aspergillosis, association with tumors of brain, lung.
		Zearalenone —Organisms: *Fusarium graminearum*, *Fusarium culmorum*, *Fusarium cerealis*, *Fusarium equiseti*, *Fusarium verticillioides*, *Fusarium incarnatum*		
16	Zearalenone	Estrogen mimic	Enters the body through the lungs, mucous membranes, or the skin	Can lead to reproductive issues such as low sperm count, inability to ovulate, spontaneous abortions. May lead to early puberty in girls

References : https://realtimelab.com/gliotoxin/; https://realtimelab.com/aflatoxins/; https://realtimelab.com/trichothecenes/; https://realtimelab.com/ochratoxins/ ; https://realtimelab.com/molds/

972.492.0419
972.243.7759

4100 Fairway Court, Suite 600
Carrollton, TX 75010

CAP #7210193 CLIA #: 45D1061736

www.RealTimeLab.com

Accompanying Laboratory Report(s)

Sample Results should be read in conjunction with and in specific reference to the accompanying Laboratory Reports.

REALTIME LABORATORIES INC
Cutting edge. Breaking barriers.

RealTime Laboratories, Inc
4100 Fairway Drive, Ste 600
Carrollton, TX 75010
Phone: 1-972-492-0419
Fax: 1-972-243-7759
Website: www.realtimelab.com
Email: info@realtimelab.com
CLIA #: 45D1061736
CAP #: 7210193
TaxId#: 45-0669342

EMMA (ENVIRONMENTAL MOLD
MYCOTOXIN ASSESSMENT)
REPORT FORM
09/18/2020

Company: ENVDAT-FLORIDA
Project: ICP Office HVAC
Location: 2710 Brantley Blvd
Naples, FL 34117
Date of Receipt: 09/16/2020
Date of Report: 09/18/2020

Accession No: EN600226EM
Date of Service: 09/12/2020
Specimen: Dust

Procedure: EMMA
TYPE: Quantitative PCR (Polymerase Chain Reaction)

Code	TEST	Results (ng of DNA/mL)	Spores/mL
EM001	Aspergillus flavus	0.0000	0
EM002	Aspergillus fumigatus	0.4927	6
EM003	Aspergillus niger	13.3148	7
EM004	Aspergillus ochraceus	0.0000	0
EM005	Aspergillus versicolor	0.0000	0
EM006	Chaetomium globosum	0.0300	1
EM010	Stachybotrys chartarum	0.0000	0
EM013	Aspergillus terreus	0.0000	0
EM014	Candida auris	0.0000	0
EM015	Fusarium solani	0.0000	0

Result Comments

COMPOSITE: HVAC RETURN SAMPLES PRE V 1

Sheri Ayus

Director Signature

RTL maintains liability limited to cost of analysis. Interpretation of the data contained in this report is the responsibility of the client. This report relates only to the samples reported above and may not be reproduced, except in full, without written approval by RTL. The above test report relates only to the items tested. RTL bears no responsibility for sample collection activities or analytical method limitations. NOTE: Results are presented as "fungal load" that measures the amount of DNA in the given sample.
For further information use the link below.
https://realtimelab.com/wp-content/uploads/2019/04/Fungal-load-EMMA-final-report-DH-Apr#-18-2019.pdf

REALTIME
LABORATORIES, INC.
Cutting edge. Breaking barriers.

**ENVIRONMENTAL MYCOTOXIN
PANEL REPORT FORM
09/21/2020**

RealTime Laboratories, Inc
4100 Fairway Drive, Ste 600
Carrollton, TX 75010
Phone: 1-972-492-0419
Fax: 1-972-243-7759
Website: www.realtimelab.com
Email: info@realtimelab.com
CLIA #: 45D1051736
CAP #: 7210193
TaxId#: 45-0669342

Company: ENVDAT-FLORIDA
Project: ICP Office HVAC
Location: 2710 Brantley Blvd
    Naples, FL 34117
Date of Receipt: 09/16/2020
Date of Report: 09/21/2020

Accession No: EN600226EM
Date of Service: 09/12/2020
Specimen: Dust

Procedure Type: Semi-quantitative procedure by ELISA
List of Mycotoxins tested in the Panel

Ochratoxin A
Aflatoxin Group: (B1, B2, G1, G2)
Trichothecene Group (Macrocyclic): Roridin A, Roridin E, Roridin H, Roridin L-2, Verrucarin A, Verrucarin J, Satratoxin G, Satratoxin H, Isosatratoxin F
Gliotoxin Derivative

Results:

Code	Test	Specimen	Value	Result	Not Present if less than	Equivocal if between	Present if greater or equal
D8501	Ochratoxin A	Dust	12.62000 ppb	Present	1.8 ppb	1.8-2.0 ppb	2.0 ppb
D8502	Aflatoxin Group: (B1, B2, G1, G2)	Dust	1.72700 ppb	Present	0.8 ppb	0.8-1.0 ppb	1.0 ppb
D8503	Trichothecene Group (Macrocyclic): Roridin A, Roridin E, Roridin H, Roridin L-2, Verrucarin A, Verrucarin J, Satratoxin G, Satratoxin H, Isosatratoxin F	Dust	0.04200 ppb	Present	0.02 ppb	0.02-0.03 ppb	0.03 ppb
D8510	Gliotoxin Derivative	Dust	0.86800 ppb	Equivocal	0.5 ppb	0.5-1.0 ppb	1.0 ppb

COMPOSITE: HVAC RETURN SAMPLES PRE V 1

Sheri Ayers

Director Signature _____

Tests such as this should be used only in conjunction with other medically established diagnostic elements (e.g.,symptoms, history, clinical impressions, results from other tests, etc). Physicians should use all the information available to them to diagnose and determine appropriate treatment for their patients.
Disclaimer: This test was developed and its performance characteristics determined by RealTime Lab. It has not been cleared or approved by the U.S. Food and Drug Administration. The FDA has determined that such clearance or approval is not necessary. This laboratory is certified under the Clinical Laboratory Improvement Amendments of 1988 (CLIA-88) as qualified to perform high complexity clinical laboratory testing.

10/12/2020   Print The Report EN600203EM-EMMA

REALTIME
LABORATORIES INC.
4100 Fairway Drive, Ste 600
Carrollton, TX 75010
www.realtimelab.com

EMMA (ENVIRONMENTAL MOLD MYCOTOXIN ASSESSMENT)
REPORT FORM
10/09/2020

COMPANY INFORMATION	ORDER INFORMATION	SAMPLE INFORMATION	LAB INFORMATION
Company: ENVDAT-FLORIDA	**Accession No:** EN600203EM	**Date of Receipt:** 10/07/2020	**Phone:** 1-972-492-0419
Project: ICP Office HVAC	**Date of Service:** 09/12/2020	**Time of Receipt:** 13:03 CDT	**Fax:** 1-972-243-7759
Location: 2710 Brantley Blvd Naples, FL 34117	**Date of Report:** 10/09/2020	**Date of Collection:** 09/12/2020	**Email:** info@realtimelab.com
Project Phone: 239-604-1228	**Contact:** David Quigley	**Time of Collection:** 00:00 CDT	**CLIA #:** 45D1051736
Project Email: NA		**Sample Type:** Dust	**CAP #:** 7210193
			Tax ID #: 0669342

Procedure: EMMA
TYPE: Quantitative PCR (Polymerase Chain Reaction)

Code	TEST	Results (ng of DNA/mL)	Spores/mL
EM001	Aspergillus flavus	0.0000	0
EM002	Aspergillus fumigatus	0.0000	0
EM003	Aspergillus niger	3.7282	2
EM004	Aspergillus ochraceus	0.0000	0
EM005	Aspergillus versicolor	0.0000	0
EM006	Chaetomium globosum	0.0000	0
EM010	Stachybotrys chartarum	0.0000	0
EM013	Aspergillus terreus	0.0000	0
EM014	Candida auris	0.0000	0
EM015	Fusarium solani	0.0000	0

Result Comments
No Comments

REPORT COMMENTS:
Composite-HVAC Air Handler samples. Prev 2 Air Handler insulation, Prev 2 Air Handler core tray, Prev 2 Air Handler core insulation.

Sheri Ayus

Director Signature _____

1/2

10/12/2020   Print The Report EN600203EM-MYCOTOXIN

REALTIME
LABORATORIES INC.
4100 Fairway Drive, Ste 600
Carrollton, TX 75010
www.realtimelab.com

ENVIRONMENTAL MYCOTOXIN PANEL REPORT FORM
10/09/2020

COMPANY INFORMATION	ORDER INFORMATION	SAMPLE INFORMATION	LAB INFORMATION
Company: ENVDAT-FLORIDA	**Accession No:** EN600203EM	**Date of Receipt:** 10/07/2020	**Phone:** 1-972-492-0419
Project: ICP Office HVAC	**Date of Service:** 09/12/2020	**Time of Receipt:** 13:03 CDT	**Fax:** 1-972-243-7759
Location: 2710 Brantley Blvd Naples, FL 34117	**Date of Report:** 10/09/2020	**Date of Collection:** 09/12/2020	**Email:** info@realtimelab.com
Project Phone: 239-604-1228	**Contact:** David Quigley	**Time of Collection:** 00:00 CDT	**CLIA #:** 45D1051736
Project Email: NA		**Sample Type:** Dust	**CAP #:** 7210193
			Tax ID #: 0669342

Procedure Type: Semi-quantitative procedure by ELISA

List of Mycotoxins tested in the Panel

Ochratoxin A
Aflatoxin Group: (B1, B2, G1, G2)
Trichothecene Group (Macrocyclic): Roridin A, Roridin E, Roridin H, Roridin L-2, Verrucarin A, Verrucarin J,
Satratoxin G, Satratoxin H, Isosatratoxin F
Gliotoxin Derivative
Zearalenone
Zearalenone Standalone

Results:

Code	Test	Specimen	Value	Result	Not Present if less than	Equivocal if between	Present if greater or equal
D8501	Ochratoxin A	Dust	0.13000 ppb	Not Present	1.8 ppb	1.8-2.0 ppb	2.0 ppb
D8502	Aflatoxin Group: (B1, B2, G1, G2)	Dust	1.28300 ppb	Present	0.8 ppb	0.8-1.0 ppb	1.0 ppb
D8503	Trichothecene Group (Macrocyclic): Roridin A, Roridin E, Roridin H, Roridin L-2, Verrucarin A, Verrucarin J, Satratoxin G, Satratoxin H, Isosatratoxin F	Dust	0.01900 ppb	Not Present	0.02 ppb	0.02-0.03 ppb	0.03 ppb
D8510	Gliotoxin Derivative	Dust	9.88400 ppb	Present	0.5 ppb	0.5-1.0 ppb	1.0 ppb
D8512	Zearalenone	Dust	0.27100 ppb	Not Present	0.5 ppb	0.5-0.7 ppb	0.7 ppb

REPORT COMMENTS:

Composite-HVAC Air Handler samples. Prev 2 Air Handler insulation, Prev 2 Air Handler core tray, Prev 2 Air Handler core insulation.

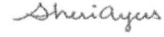

Director Signature _____

1/2

9/18/2020   Print The Report EN600229EM-EMMA

RealTime Laboratories, Inc
4100 Fairway Drive, Ste 600
Carrollton, TX 75010
Phone: 1-972-492-0419
Fax: 1-972-243-7759
Website: www.realtimelab.com
Email: info@realtimelab.com
CLIA #: 45D1051736
CAP #: 7210193
TaxId#: 45-0669342

EMMA (ENVIRONMENTAL MOLD
MYCOTOXIN ASSESSMENT)
REPORT FORM
09/18/2020

Company: ENVDAT-FLORIDA
Project: ICP Office HVAC
Location: 2710 Brantley Blvd
      Naples, FL 34117
Date of Receipt: 09/16/2020
Date of Report: 09/18/2020

Accession No: EN600229EM
Date of Service: 09/12/2020
Specimen: Dust

Procedure: EMMA
TYPE: Quantitative PCR (Polymerase Chain Reaction)

Code	TEST	Results (ng of DNA/mL)	Spores/mL
EM001	Aspergillus flavus	0.0000	0
EM002	Aspergillus fumigatus	0.0000	0
EM003	Aspergillus niger	0.0000	0
EM004	Aspergillus ochraceus	0.0000	0
EM005	Aspergillus versicolor	0.0000	0
EM006	Chaetomium globosum	0.0000	0
EM010	Stachybotrys chartarum	0.0000	0
EM013	Aspergillus terreus	0.0000	0
EM014	Candida auris	0.0000	0
EM015	Fusarium solani	0.0000	0

Result Comments

Composite HVAC supply samples: PREV 3 SUPPLY PLENUM, PREV 3 SUPPLY VENT 1, PREV 3 SUPPLY VENT 2

Sheri Ayers

Director Signature

1/2

9/18/2020   Print The Report EN600229EM-MYCOTOXIN

RealTime Laboratories, Inc
4100 Fairway Drive, Ste 600
Carrollton, TX 75010
Phone: 1-972-492-0419
ENVIRONMENTAL MYCOTOXIN   Fax: 1-972-243-7759
PANEL REPORT FORM   Website: www.realtimelab.com
09/17/2020   Email: info@realtimelab.com
                                                           CLIA #: 45D1051736
                                                           CAP #: 7210193
                                                           TaxId#: 45-0669342

Company: ENVDAT-FLORIDA
Project: ICP Office HVAC
Location: 2710 Brantley Blvd   **Accession No:** EN600229EM
           Naples, FL 34117   **Date of Service:** 09/12/2020
Date of Receipt: 09/16/2020   **Specimen:** Dust
Date of Report: 09/17/2020

Procedure Type: Semi-quantitative procedure by ELISA
List of Mycotoxins tested in the Panel

Ochratoxin A
Aflatoxin Group: (B1, B2, G1, G2)
Trichothecene Group (Macrocyclic): Roridin A, Roridin E, Roridin H, Roridin L-2, Verrucarin A, Verrucarin J, Satratoxin G, Satratoxin H, Isosatratoxin F
Gliotoxin Derivative

Results:

Code	Test	Specimen	Value	Result	Not Present if less than	Equivocal if between	Present if greater or equal
D8501	Ochratoxin A	Dust	0.54500 ppb	Not Present	1.8 ppb	1.8-2.0 ppb	2.0 ppb
D8502	Aflatoxin Group: (B1, B2, G1, G2)	Dust	0.32500 ppb	Not Present	0.8 ppb	0.8-1.0 ppb	1.0 ppb
D8503	Trichothecene Group (Macrocyclic): Roridin A, Roridin E, Roridin H, Roridin L-2, Verrucarin A, Verrucarin J, Satratoxin G, Satratoxin H, Isosatratoxin F	Dust	0.00000 ppb	Not Present	0.02 ppb	0.02-0.03 ppb	0.03 ppb
D8510	Gliotoxin Derivative	Dust	0.24500 ppb	Not Present	0.5 ppb	0.5-1.0 ppb	1.0 ppb

Composite HVAC supply samples: PREV 3 SUPPLY PLENUM, PREV 3 SUPPLY VENT 1, PREV 3 SUPPLY VENT 2

Sheri Ayers

Director Signature _____

Tests such as this should be used only in conjunction with other medically established diagnostic elements (e.g.,symptoms, history, clinical impressions, results from other tests, etc). Physicians should use all the information available to them to diagnose and determine appropriate treatment for their patients.
Disclaimer: This test was developed and its performance characteristics determined by RealTime Lab. It has not been cleared or approved by the U.S. Food and Drug Administration. The FDA has determined that such clearance or approval is not necessary. This laboratory is certified under the Clinical Laboratory Improvement Amendments of 1988 (CLIA-88) as qualified to perform high complexity clinical laboratory testing.

1/2

9/23/2020   Print The Report EN600216EM-EMMA

RealTime Laboratories, Inc
4100 Fairway Drive, Ste 600
Carrollton, TX 75010
Phone: 1-972-492-0419
Fax: 1-972-243-7759
Website: www.realtimelab.com
Email: info@realtimelab.com
CLIA #: 45D1051736
CAP #: 7210193
TaxId#: 45-0669342

**EMMA (ENVIRONMENTAL MOLD MYCOTOXIN ASSESSMENT)
REPORT FORM
09/23/2020**

Company: ENVDAT-FLORIDA
Project: ICP Office HVAC
Location: 2710 Brantley Blvd   **Accession No:** EN600216EM
             Naples, FL 34117   **Date of Service:** 09/16/2020
Date of Receipt: 09/21/2020   **Specimen:** Dust
Date of Report: 09/23/2020

Procedure: EMMA
TYPE: Quantitative PCR (Polymerase Chain Reaction)

Code	TEST	Results (ng of DNA/mL)	Spores/mL
EM001	Aspergillus flavus	0.0000	0
EM002	Aspergillus fumigatus	0.0000	0
EM003	Aspergillus niger	0.2451	1
EM004	Aspergillus ochraceus	0.0000	0
EM005	Aspergillus versicolor	0.0000	0
EM006	Chaetomium globosum	0.0000	0
EM010	Stachybotrys chartarum	0.0000	0
EM013	Aspergillus terreus	0.0000	0
EM014	Candida auris	0.0000	0
EM015	Fusarium solani	0.0000	0

Result Comments

Composite HVAC: Post V 4

Sheri Ayers

Director Signature _____

RTL maintains liability limited to cost of analysis. Interpretation of the data contained in this report is the responsibility of the client. This report relates only to the samples reported above and may not
be reproduced, except in full, without written approval by RTL. The above test report relates only to the items tested. RTL bears no responsibility for sample collection activities or analytical method
limitations.
NOTE: Results are presented as "fungal load" that measures the amount of DNA in the given sample.
For further information use the link below.
https://realtimelab.com/wp-content/uploads/2019/04/Fungal-load-EMMA-final-report-DH-April-18-2019.pdf

1/2

9/23/2020   Print The Report EN600216EM-MYCOTOXIN

REALTIME
LABORATORIES, INC.
Cutting edge. Breaking barriers.

RealTime Laboratories, Inc
4100 Fairway Drive, Ste 600
Carrollton, TX 75010
Phone: 1-972-492-0419
Fax: 1-972-243-7759
Website: www.realtimelab.com
Email: info@realtimelab.com
CLIA #: 45D1051736
CAP #: 7210193
TaxId#: 45-0669342

ENVIRONMENTAL MYCOTOXIN PANEL REPORT FORM
09/22/2020

Company: ENVDAT-FLORIDA
Project: ICP Office HVAC
Location: 2710 Brantley Blvd
           Naples, FL 34117
Date of Receipt: 09/21/2020
Date of Report: 09/22/2020

Accession No: EN600216EM
Date of Service: 09/16/2020
Specimen: Dust

Procedure Type: Semi-quantitative procedure by ELISA
List of Mycotoxins tested in the Panel

Ochratoxin A
Aflatoxin Group: (B1, B2, G1, G2)
Trichothecene Group (Macrocyclic): Roridin A, Roridin E, Roridin H, Roridin L-2, Verrucarin A, Verrucarin J, Satratoxin G, Satratoxin H, Isosatratoxin F
Gliotoxin Derivative

Results:

Code	Test	Specimen	Value	Result	Not Present if less than	Equivocal if between	Present if greater or equal
D8501	Ochratoxin A	Dust	0.04700 ppb	Not Present	1.8 ppb	1.8-2.0 ppb	2.0 ppb
D8502	Aflatoxin Group: (B1, B2, G1, G2)	Dust	0.41600 ppb	Not Present	0.8 ppb	0.8-1.0 ppb	1.0 ppb
D8503	Trichothecene Group (Macrocyclic): Roridin A, Roridin E, Roridin H, Roridin L-2, Verrucarin A, Verrucarin J, Satratoxin G, Satratoxin H, Isosatratoxin F	Dust	0.00700 ppb	Not Present	0.02 ppb	0.02-0.03 ppb	0.03 ppb
D8510	Gliotoxin Derivative	Dust	0.08200 ppb	Not Present	0.5 ppb	0.5-1.0 ppb	1.0 ppb

Composite HVAC: Post V 4

Sheri Ayers

Director Signature _____

Tests such as this should be used only in conjunction with other medically established diagnostic elements (e.g.,symptoms, history, clinical impressions, results from other tests, etc). Physicians should use all the information available to them to diagnose and determine appropriate treatment for their patients.
Disclaimer: This test was developed and its performance characteristics determined by RealTime Lab. It has not been cleared or approved by the U.S. Food and Drug Administration. The FDA has determined that such clearance or approval is not necessary. This laboratory is certified under the Clinical Laboratory Improvement Amendments of 1988 (CLIA-88) as qualified to perform high complexity clinical laboratory testing.

1/2

9/23/2020   Print The Report EN600227EM-EMMA

RealTime Laboratories, Inc
4100 Fairway Drive, Ste 600
Carrollton, TX 75010
Phone: 1-972-492-0419
Fax: 1-972-243-7759
Website: www.realtimelab.com
Email: info@realtimelab.com
CLIA #: 45D1051736
CAP #: 7210193
Taxid#: 45-0669342

**EMMA (ENVIRONMENTAL MOLD MYCOTOXIN ASSESSMENT)
REPORT FORM
09/23/2020**

Company: ENVDAT-FLORIDA
Project: ICP Office HVAC
Location: 2710 Brantley Blvd
             Naples, FL 34117
Date of Receipt: 09/21/2020
Date of Report: 09/23/2020

Accession No: EN600227EM
Date of Service: 09/16/2020
Specimen: Dust

Procedure: EMMA
TYPE: Quantitative PCR (Polymerase Chain Reaction)

Code	TEST	Results (ng of DNA/mL)	Spores/mL
EM001	Aspergillus flavus	0.0000	0
EM002	Aspergillus fumigatus	0.0000	0
EM003	Aspergillus niger	0.0000	0
EM004	Aspergillus ochraceus	0.0000	0
EM005	Aspergillus versicolor	0.0000	0
EM006	Chaetomium globosum	0.0000	0
EM010	Stachybotrys chartarum	0.0000	0
EM013	Aspergillus terreus	0.0000	0
EM014	Candida auris	0.0000	0
EM015	Fusarium solani	0.0000	0

Result Comments

Composite HVAC: Post V 5

Director Signature _____

RTL maintains liability limited to cost of analysis. Interpretation of the data contained in this report is the responsibility of the client. This report relates only to the samples reported above and may not be reproduced, except in full, without written approval by RTL. The above test report relates only to the items tested. RTL bears no responsibility for sample collection activities or analytical method limitations.
NOTE: Results are presented as "fungal load" that measures the amount of DNA in the given sample.
For further information use the link below.
https://realtimelab.com/wp-content/uploads/2019/04/Fungal-load-EMMA-final-report-DH-April-18-2019.pdf

1/2

9/23/2020   Print The Report EN600227EM-MYCOTOXIN

RealTime Laboratories, Inc
4100 Fairway Drive, Ste 600
Carrollton, TX 75010
Phone: 1-972-492-0419
Fax: 1-972-243-7759
Website: www.realtimelab.com
Email: info@realtimelab.com
CLIA #: 45D1051736
CAP #: 7210193
TaxId#: 45-0669342

Cutting edge. Breaking barriers.

ENVIRONMENTAL MYCOTOXIN PANEL REPORT FORM
09/22/2020

Company: ENVDAT-FLORIDA
Project: ICP Office HVAC
Location: 2710 Brantley Blvd
           Naples, FL 34117
Date of Receipt: 09/21/2020
Date of Report: 09/22/2020

Accession No: EN600227EM
Date of Service: 09/16/2020
Specimen: Dust

Procedure Type: Semi-quantitative procedure by ELISA
List of Mycotoxins tested in the Panel

Ochratoxin A
Aflatoxin Group: (B1, B2, G1, G2)
Trichothecene Group (Macrocyclic): Roridin A, Roridin E, Roridin H, Roridin L-2, Verrucarin A, Verrucarin J, Satratoxin G, Satratoxin H, Isosatratoxin F
Gliotoxin Derivative

Results:

Code	Test	Specimen	Value	Result	Not Present if less than	Equivocal if between	Present if greater or equal
D8501	Ochratoxin A	Dust	0.05000 ppb	Not Present	1.8 ppb	1.8-2.0 ppb	2.0 ppb
D8502	Aflatoxin Group: (B1, B2, G1, G2)	Dust	0.32500 ppb	Not Present	0.8 ppb	0.8-1.0 ppb	1.0 ppb
D8503	Trichothecene Group (Macrocyclic): Roridin A, Roridin E, Roridin H, Roridin L-2, Verrucarin A, Verrucarin J, Satratoxin G, Satratoxin H, Isosatratoxin F	Dust	0.01700 ppb	Not Present	0.02 ppb	0.02-0.03 ppb	0.03 ppb
D8510	Gliotoxin Derivative	Dust	0.04900 ppb	Not Present	0.5 ppb	0.5-1.0 ppb	1.0 ppb

Composite HVAC: Post V 5

Sheri Ayers

Director Signature _____

Tests such as this should be used only in conjunction with other medically established diagnostic elements (e.g.,symptoms, history, clinical impressions, results from other tests, etc.). Physicians should use all the information available to them to diagnose and determine appropriate treatment for their patients.
Disclaimer: This test was developed and its performance characteristics determined by RealTime Lab. It has not been cleared or approved by the U.S. Food and Drug Administration. The FDA has determined that such clearance or approval is not necessary. This laboratory is certified under the Clinical Laboratory Improvement Amendments of 1988 (CLIA-88) as qualified to perform high complexity clinical laboratory testing.

1/2

9/23/2020   Print The Report EN600228EM-EMMA

RealTime Laboratories, Inc
4100 Fairway Drive, Ste 600
Carrollton, TX 75010
Phone: 1-972-492-0419
Fax: 1-972-243-7759
Website: www.realtimelab.com
Email: info@realtimelab.com
CLIA #: 45D1051736
CAP #: 7210193
TaxId#: 45-0669342

EMMA (ENVIRONMENTAL MOLD MYCOTOXIN ASSESSMENT)
REPORT FORM
09/23/2020

Company: ENVDAT-FLORIDA
Project: ICP Office HVAC
Location: 2710 Brantley Blvd
Naples, FL 34117
Date of Receipt: 09/21/2020
Date of Report: 09/23/2020

Accession No: EN600228EM
Date of Service: 09/16/2020
Specimen: Dust

Procedure: EMMA
TYPE: Quantitative PCR (Polymerase Chain Reaction)

Code	TEST	Results (ng of DNA/mL)	Spores/mL
EM001	Aspergillus flavus	0.0000	0
EM002	Aspergillus fumigatus	0.0000	0
EM003	Aspergillus niger	0.0000	0
EM004	Aspergillus ochraceus	0.0000	0
EM005	Aspergillus versicolor	0.0000	0
EM006	Chaetomium globosum	0.0000	0
EM010	Stachybotrys chartarum	0.0000	0
EM013	Aspergillus terreus	0.0000	0
EM014	Candida auris	0.0000	0
EM015	Fusarium solani	0.0000	0

Result Comments

Composite HVAC: Post V 6

Sheri Ayers

Director Signature _____

1/2

9/23/2020   Print The Report EN600228EM-MYCOTOXIN

RealTime Laboratories, Inc
4100 Fairway Drive, Ste 600
Carrollton, TX 75010
Phone: 1-972-492-0419
Fax: 1-972-243-7759
Website: www.realtimelab.com
Email: info@realtimelab.com
CLIA #: 45D1051736
CAP #: 7210193
TaxId#: 45-0669342

ENVIRONMENTAL MYCOTOXIN PANEL REPORT FORM
09/22/2020

Company: ENVDAT-FLORIDA
Project: ICP Office HVAC
Location: 2710 Brantley Blvd   **Accession No:** EN600228EM
        Naples, FL 34117   **Date of Service:** 09/16/2020
Date of Receipt: 09/21/2020   **Specimen:** Dust
Date of Report: 09/22/2020

Procedure Type: Semi-quantitative procedure by ELISA
List of Mycotoxins tested in the Panel

Ochratoxin A
Aflatoxin Group: (B1, B2, G1, G2)
Trichothecene Group (Macrocyclic): Roridin A, Roridin E, Roridin H, Roridin L-2, Verrucarin A, Verrucarin J, Satratoxin G, Satratoxin H, Isosatratoxin F
Gliotoxin Derivative

Results:

Code	Test	Specimen	Value	Result	Not Present if less than	Equivocal if between	Present if greater or equal
D8501	Ochratoxin A	Dust	0.03500 ppb	Not Present	1.8 ppb	1.8-2.0 ppb	2.0 ppb
D8502	Aflatoxin Group: (B1, B2, G1, G2)	Dust	0.32800 ppb	Not Present	0.8 ppb	0.8-1.0 ppb	1.0 ppb
D8503	Trichothecene Group (Macrocyclic): Roridin A, Roridin E, Roridin H, Roridin L-2, Verrucarin A, Verrucarin J, Satratoxin G, Satratoxin H, Isosatratoxin F	Dust	0.00000 ppb	Not Present	0.02 ppb	0.02-0.03 ppb	0.03 ppb
D8510	Gliotoxin Derivative	Dust	0.06600 ppb	Not Present	0.5 ppb	0.5-1.0 ppb	1.0 ppb

Composite HVAC: Post V 6

Director Signature _____

Tests such as this should be used only in conjunction with other medically established diagnostic elements (e.g.,symptoms, history, clinical impressions, results from other tests, etc). Physicians should use all the information available to them to diagnose and determine appropriate treatment for their patients.
Disclaimer: This test was developed and its performance characteristics determined by RealTime Lab. It has not been cleared or approved by the U.S. Food and Drug Administration. The FDA has determined that such clearance or approval is not necessary. This laboratory is certified under the Clinical Laboratory Improvement Amendments of 1988 (CLIA-88) as qualified to perform high complexity clinical laboratory testing.

1/2

REALTIME
LABORATORIES INC.
4100 Fairway Drive, Ste 600
Carrollton, TX 75010
www.realtimelab.com

EMMA (ENVIRONMENTAL MOLD MYCOTOXIN ASSESSMENT)
REPORT FORM
10/13/2020

COMPANY INFORMATION	ORDER INFORMATION	SAMPLE INFORMATION	LAB INFORMATION
Company: ENVDAT-FLORIDA	**Accession No:** EN600092EM	**Date of Receipt:** 10/09/2020	**Phone:** 1-972-492-0419
Project: ICP Office I HVAC	**Date of Service:** 10/07/2020	**Time of Receipt:** 16:17 CDT	**Fax:** 1-972-243-7759
Location: 2710 Brantley Blvd Naples, FL 34117	**Date of Report:** 10/13/2020	**Date of Collection:** 10/07/2020	**Email:** info@realtimelab.com
Project Phone: 239-604-1228	**Contact:** David Quigley	**Time of Collection:** 00:00 CDT	**CLIA #:** 45D1051736
Project Email: NA		**Sample Type:** Dust	**CAP #:** 7210193
			Tax ID #: 0669342

Procedure: EMMA

TYPE: Quantitative PCR (Polymerase Chain Reaction)

Code	TEST	Results (ng of DNA/mL)	Spores/mL
EM001	Aspergillus flavus	0.0000	0
EM002	Aspergillus fumigatus	0.0000	0
EM003	Aspergillus niger	0.0000	0
EM004	Aspergillus ochraceus	0.0000	0
EM005	Aspergillus versicolor	0.0000	0
EM006	Chaetomium globosum	0.0000	0
EM010	Stachybotrys chartarum	0.0000	0
EM013	Aspergillus terreus	0.0000	0
EM014	Candida auris	0.0000	0
EM015	Fusarium solani	0.0000	0

Result Comments
No Comments

REPORT COMMENTS:

Post V + 1 HVAC Floor Returns; Post V + 1 HVAC Ceiling Returns; Post V + 1 HVAC Return Plenum

Sheri Ayers

Director Signature _____

1/2

10/14/2020   Print The Report EN600092EM-MYCOTOXIN

 **REALTIME** LABORATORIES INC.

4100 Fairway Drive, Ste 600
Carrollton, TX 75010
www.realtimelab.com

ENVIRONMENTAL MYCOTOXIN PANEL REPORT FORM
10/13/2020

COMPANY INFORMATION

Company: ENVDAT-FLORIDA
Project: ICP Office I HVAC
Location: 2710 Brantley Blvd Naples, FL 34117
Project Phone: 239-604-1228
Project Email: NA

ORDER INFORMATION

Accession No: EN600092EM
Date of Service: 10/07/2020
Date of Report: 10/13/2020
Contact: David Quigley

SAMPLE INFORMATION

Date of Receipt: 10/09/2020
Time of Receipt: 16:17 CDT
Date of Collection: 10/07/2020
Time of Collection: 00:00 CDT
Sample Type: Dust

LAB INFORMATION

Phone: 1-972-492-0419
Fax: 1-972-243-7759
Email: info@realtimelab.com
CLIA #: 45D1051736
CAP #: 7210193
Tax ID #: 0669342

Procedure Type: Semi-quantitative procedure by ELISA

List of Mycotoxins tested in the Panel
Ochratoxin A
Aflatoxin Group: (B1, B2, G1, G2)
Trichothecene Group (Macrocyclic): Roridin A, Roridin E, Roridin H, Roridin L-2, Verrucarin A, Verrucarin J, Satratoxin G, Satratoxin H, Isosatratoxin F
Gliotoxin Derivative
Zearalenone
Zearalenone Standalone

Results:

Code	Test	Specimen	Value	Result	Not Present if less than	Equivocal if between	Present if greater or equal
D8501	Ochratoxin A	Dust	0.65700 ppb	Not Present	1.8 ppb	1.8-2.0 ppb	2.0 ppb
D8502	Aflatoxin Group: (B1, B2, G1, G2)	Dust	0.29800 ppb	Not Present	0.8 ppb	0.8-1.0 ppb	1.0 ppb
D8503	Trichothecene Group (Macrocyclic): Roridin A, Roridin E, Roridin H, Roridin L-2, Verrucarin A, Verrucarin J, Satratoxin G, Satratoxin H, Isosatratoxin F	Dust	0.01000 ppb	Not Present	0.02 ppb	0.02-0.03 ppb	0.03 ppb
D8510	Gliotoxin Derivative	Dust	0.13400 ppb	Not Present	0.5 ppb	0.5-1.0 ppb	1.0 ppb
D8512	Zearalenone	Dust	0.31800 ppb	Not Present	0.5 ppb	0.5-0 7 ppb	0.7 ppb

REPORT COMMENTS:
Post V + 1 HVAC Floor Returns; Post V + 1 HVAC Ceiling Returns; Post V + 1 HVAC Return Plenum

Director Signature _____

Tests such as this should be used only in conjunction with other medically established diagnostic elements (e.g.,symptoms, history, clinical impressions, results from other tests, etc). Physicians should use all the information available to them to diagnose and determine appropriate treatment for their patients.
Disclaimer: This test was developed and its performance characteristics determined by RealTime Lab. It has not been cleared or approved by the U.S. Food and Drug Administration. The FDA has determined that such clearance or approval is not necessary. This laboratory is certified under the Clinical Laboratory Improvement Amendments of 1988 (CLIA-88) as qualified to perform high complexity clinical laboratory testing.

1/2

ABOUT THE AUTHOR

A native New Zealander, David Mark Quigley has carved out a niche for himself in International Hazardous and Biohazard Substances. In the US, he specializes in indoor mycotoxin contamination: the consultation, assessment, identification, treatment, and validation of these biohazardous toxins.

His latest book, *Mycotoxin Deactivation: A Successful Mycotoxin Treatment and Reduction Case Study*, describes the radical solution he has discovered to the previously unsolved problem of in situ deactivation. The old method required complete removal and replacement of all contaminated building materials, fabrics, and mechanical equipment. His case study documents the successful deactivation of indoor mycotoxin contamination. This innovative approach is a revolutionary advancement and a game-changing process. When adopted, this will be tremendously beneficial across multiple environments and many different industries.

He lives in Naples, Florida, with his wife and numerous furry freeloaders, in a home he built in his spare time. He is obsessed with animals and nature and has chased adventure across Europe, Australia, and Africa. Inspired by his extensive travels, he decided to tackle his dyslexia by writing a novel, *Scars of the Leopard* and unexpectedly discovered a love of writing. He is the author of three further action adventures, *The Last Rhino, White Gold*, and *African Lion.* By purchasing his books, you are seamlessly donating to wildlife conservation, as a percentage of every sale is given via his non-profit foundation.

NONFICTION WORKS BY AUTHOR

Mycotoxin Treatment Series:

Mycotoxin Deactivation:
A Successful Mycotoxin Treatment and Reduction Case Study,

FICTION WORKS BY AUTHOR

African Series:

Series Prequel – Scars of the Leopard
https://geni.us/Mycotoxin_Scars

Book 1 – The Last Rhino:
https://geni.us/Mycotoxin_LastRhino

Book 2 - White Gold:
https://geni.us/Mycotoxin_WhiteGold

Book 3 - African Lion:
https://geni.us/Mycotoxin_AfricanLion

CONTACT ME FOR
MY NONFICTION WORK

For more information about myself and my company BioRisk
please visit the link below:
https://biorisk.us/

If you would like to contact me personally please email or call the
number below. I would look forward to hearing from you:
info@biorisk.us
(877) BioRisk

CONTACT ME FOR
MY FICTION WORK

Get personalized book information and up-to-date news about my works:
https://davidmarkquigley.com/

Be the first to know about new releases, awesome giveaways and news by signing up for the VIP mailing list:
books@davidmarkquigley.com

REVIEWS

Did you find this book informative? If so, I would love to hear about it. Honest reviews help readers find the right book for their needs, especially when faced with the vexing problem of treating and deactivating mycotoxins. To leave a review, please head to *Mycotoxin Deactivation: A Successful Mycotoxin Treatment and Reduction Case Study's*, Amazon page, scroll to the bottom of the page under "More about the author", and select "Write a customer review".

Or alternately, please use the link below that will take you straight to the "Write a customer review" section: https://www.amazon.com/product-reviews/B09XN6PBM7

I hope you enjoyed reading this book as much as I did writing it. It has been a pleasure having you as one of my readers. Thank you!

Did you know that a portion of all my fiction proceeds is seamlessly donated to wildlife causes around the world?
To find out more
To donate: https://davidmarkquigley.com/wildlife-foundation

Printed in Great Britain
by Amazon

39829757R00066